高水平地方应用型大学建设系列教材

环境工程专业英语

郑红艾　蒋路漫　时鹏辉　编著

北　京
冶 金 工 业 出 版 社
2024

内 容 提 要

本书分为上、下两篇，共3章。上篇为理论知识，主要内容包括专业英语概述和专业英语翻译；下篇为英语实例，主要内容包括水污染及控制技术、大气污染控制技术和环境影响等。

本书可作为高等院校环境科学、环境工程专业的英语教材，也可供从事环境科学工作的相关人员参考。

图书在版编目(CIP)数据

环境工程专业英语/郑红艾，蒋路漫，时鹏辉编著.—北京：冶金工业出版社，2022.6（2024.8重印）

高水平地方应用型大学建设系列教材

ISBN 978-7-5024-9125-3

Ⅰ.①环… Ⅱ.①郑… ②蒋… ③时… Ⅲ.①环境工程—英语—高等学校—教材 Ⅳ.①X5

中国版本图书馆CIP数据核字(2022)第064599号

环境工程专业英语

出版发行	冶金工业出版社	**电　话**	(010)64027926
地　址	北京市东城区嵩祝院北巷39号	**邮　编**	100009
网　址	www.mip1953.com	**电子信箱**	service@mip1953.com

责任编辑　刘林烨　程志宏　美术编辑　彭子赫　版式设计　郑小利

责任校对　李　娜　责任印制　禹　蕊

北京捷迅佳彩印刷有限公司印刷

2022年6月第1版，2024年8月第2次印刷

710mm×1000mm　1/16；6.75印张；71千字；98页

定价29.00元

投稿电话　(010)64027932　投稿信箱　tougao@cnmip.com.cn

营销中心电话　(010)64044283

冶金工业出版社天猫旗舰店　yjgycbs.tmall.com

（本书如有印装质量问题，本社营销中心负责退换）

《高水平地方应用型大学建设系列教材》
编　委　会

《高水平地方应用型大学建设系列教材》序

应用型大学教育是高等教育结构中的重要组成部分。高水平地方应用型高校在培养复合型人才、服务地方经济发展以及为现代产业体系提供高素质应用型人才方面越来越显现出不可替代的作用。2019年，上海电力大学获批上海市首个高水平地方应用型高校建设试点单位，为学校以能源电力为特色，着力发展清洁安全发电、智能电网和智慧能源管理三大学科，打造专业品牌，增强科研层级，提升专业水平和服务能力提出了更高的要求和发展的动力。清洁安全发电学科汇聚化学工程与工艺、材料科学与工程、材料化学、环境工程、应用化学、新能源科学与工程、能源与动力工程等专业，力求培养出具有创新意识、创新性思维和创新能力的高水平应用型建设者，为煤清洁燃烧和高效利用、水质安全与控制、环境保护、设备安全、新能源开发、储能系统、分布式能源系统等产业，输出合格应用型优秀人才，支撑国家和地方先进电力事业的发展。

教材建设是搞好应用型特色高校建设非常重要的方面。以往应用型大学的本科教学主要使用普通高等教育教学用书，实践证明并不适应在应用型高校教学使用。由于密切结合行业特色及新的生产工艺以及与先进教学实验设备相适应且实践性强的教材稀缺，迫切需要教材改革和创新。编写应用性和实践性强及有行业特色教材，是提高应用型人才培养质量的重要保障。国外一些教育发达国家的基础课教材涉及内容广、应用性强，确实值得我国应用型高校教材编写出版借鉴和参考。

为此，上海电力大学和冶金工业出版社合作共同组织了高水平地方应用型大学建设系列教材的编写，包括课程设计、实践与实习指导、实验指导等各类型的教学用书，首批出版教材 18 种。教材的编写将遵循应用型高校教学特色、学以致用、实践教学的原则，既保证教学内容的完整性、基础性，又强调其应用性，突出产教融合，将教学和学生专业知识和素质能力提升相结合。

本系列教材的出版发行，对于我校高水平地方应用型大学的建设、高素质应用型人才培养具有十分重要的现实意义，也将为教育综合改革提供示范素材。

上海电力大学校长　李和兴

2020 年 4 月

前　　言

“环境工程专业英语”是高等院校环境科学与工程专业教学的重要组成部分。随着大学教育进入国际化发展时期，专业英语的教学已成为专业应用型人才培养不可或缺的一部分。本书的目的是对本科生在专业内容方面进行英语阅读写作的系统训练，以提高学生的综合素质。

本书根据目前高等院校环境类专业英语学习的要求，系统地介绍了环境工程专业英语的词语构成、句式特点及翻译方法，并收集总结了相关领域新的研究及进展。本书旨在为环境类专业的学生提供一本比较系统的专业英语教学用书，提高学生正确获得该专业的科技信息、快速地阅读英语科技文献的能力，使学生了解专业英语特点，并初步学会专业英语写作与概括，掌握一定数量科技词汇及习惯用法，把学生学到的基础英语进行专业化训练，为社会经济发展对环境工程专业技术人才的需求奠定基础。

本书由郑红艾、蒋路漫和时鹏辉编著。本书题材较广，从纵横两个方向覆盖环境科学与工程专业的相关内容。

由于作者水平有限，书中不妥之处，敬请广大读者批评指正。

作　者

2021年11月

目　录

上篇　理论知识

下篇 英语实例

上 篇

理论知识

LILUN ZHISHI

1　专业英语概述

1.1　专业英语的特点

1.1.1　专业英语的词汇特点

专业英语的词汇特点如下。

（1）专业词汇出现的频率低。专业英语在用词方面，除了使用专业术语和行业惯用语外，大量应用的则是普通的科技词汇，用得最多的还是普通词汇。

（2）词义专一。在普通英语中一词多义和一义多词的现象极为普遍，但在专业英语中这种现象却很少见。

（3）环境工程专业广泛使用缩写词。

COD（chemical oxygen demand）　化学需氧量
BOD（biochemical oxygen demand）　生化需氧量
TOC（total organic carbon）　总有机碳
DO（dissolved oxygen）　溶解氧
POPS（persistent organic pollutants）　持久性有机污染物
TSP（total suspended particle）　总悬浮颗粒
TKN（total Kjeldahl nitrogen）　总凯氏氮

UASB（up-flow anaerobic sludge blanket） 上流式厌氧污泥床
MBR（membrane bioreactor） 膜生物反应器
SBR（sequencing batch reactor） 间歇式活性污泥法

1.1.2 专业英语的句法特点

专业英语的句法特点如下。

（1）广泛使用陈述句。专业文章主要讲述的是事实、规律、原理和做法等，因此陈述句一般用得较多。

（2）广泛使用被动语态和无主句英语。与汉语相比，专业英语具有以下三大特点：

1）冠词的使用；

2）引导词 it 和 there 的使用；

3）被动语态的广泛使用。

被动语态在专业英语中的使用比在其他场合更为广泛，主要是因为被动句比主动句更能说明需要论证的对象，更能使其位置鲜明、突出。

（3）简略表达多。专业文章要求文字简洁，因而文中的简略表达用得较多。

（4）长句使用多。专业英语描写的是专业方面的知识，注重事实和逻辑推导，所给出的定义、定律、描述的概念或叙述的生产工艺过程等都必须严谨、精确。为此，专业英语中的长句使用得较多，并且在这些长句中常常是一个主句带若干从句，从句带短语，短语带从句，相互依附，相互制约，使句子显得很长。

1.1.3 专业英语的修辞特点

专业英语的修辞特点如下。

（1）时态运用有限。在专业英语中，叙述过去的研究常用过去时（与现在不发生联系），有时也用现在完成时（与现在有直接联系，并对目前有重要影响）；讨论研究项目所基于的理论用将来时，论述理论部分用现在时。

（2）修辞手法较为单调。与文学作品相比，专业文章语言朴实，语法规范，文字讲求准确、明了，很少使用带感情色彩和华丽的辞藻，也很少使用比喻、夸张等修辞手段。

（3）逻辑性语法词使用普遍。逻辑性语法词在专业文章中使用非常普遍，可将其分类如下。

1）表示原因的词，包括：because，because of，due to，owing to，as，as a result of，caused by，for。

2）表示语气转折的词，包括：but，however，nevertheless，yet，otherwise。

3）表示逻辑、顺理连接的词，包括：so，thus，therefore，furthermore，moreover，in addition to。

4）表示限制的词，包括：only，if only，except，besides，unless。

5）表示假设的词，包括：suppose，supposing，assuming，provided，providing。

1.2　翻译的基本知识

1.2.1　翻译标准

翻译就是把一种语言文字的意义用另一种语言文字表达出来，即“give the sense of（word，speech，passage，book，eta）in another language”。翻译标准是翻译实践的准绳和衡量译文好坏的尺度，清末的严复在翻译《天演论》时，提出了著名的“信、达、雅”的翻译标准。所谓的“信”是指译文要忠实于原文；所谓的“达”是指译文要通顺明了；所谓的“雅”是指译文要讲究文字修饰，译笔要有文采。

结合专业英语的翻译特点，在进行专业英语翻译时要坚持以下两条标准。

（1）忠实：译文必须忠实、正确地传达原文的内容。

（2）通顺：译文的语言必须规范、流畅、通俗易懂。

这里没有提出“雅”作为翻译的标准之一，绝不意味着在专业文章的翻译中没有“雅”字的容身之地，而是寓“雅”于通顺之中。译文规范、流畅、通俗易懂本身就谓之修辞，意味着“雅”。

1.2.2　翻译的过程

翻译既要忠实于原作，不能随意乱译，又要选择相应而恰当的表达手段，不能机械地死译。因此，翻译应包括理解和表达这两个过程。就专业英语的翻译而论，理解就是透彻地了解，懂得、掌握原文的内容和实质；表达就是准确地运用各种不同的翻译技巧，以规范的汉语体现原文的内容。

在翻译的两个过程中，理解是第一位的，表达是第二位的。没有对原文的深刻理解，翻译就无从下手。对原文理解得不透彻，译文就会不通顺。因此，理解和表达是翻译中两个有机的联系过程，忽视哪一个都是失败的因子。

1.2.2.1 翻译的理解过程

翻译的理解过程如下。

（1）通读全文，领会大意。不少人都有个通病，就是在翻译一篇文章或一本论著之前，很少通读一遍，只顾提笔行文，进行英汉等值翻译，对号入座。译完之后连译者本人也不知译文所云，这种草率从事的态度是不可取的。在翻译文章之前，应通读全文，领略大意，这样就便于翻译时正确地选择词义，便于行文造句。

（2）明辨语法，弄清关系。译文表达不清，语义含混的一个重要原因就是没有弄清语法关系。要把每一句译得贴切，必须弄清句子内在的语法关系。

（3）结合上下文，推敲词义要透彻地理解原文，必须结合上下文，推敲词义。专业文章具有较强的科学性和逻辑性，因此词与词之间、段与段之间总是相互联系、相互制约的。孤立地、片面地理解词义，必定会出差错。因此，在推敲词义时，必须做到词不离句，句不离文，为求理解，必须做到文中求句，句中求词，以便忠实、通顺地翻译文章。

1.2.2.2 翻译的表达过程

翻译的表达过程如下。

（1）一稿初译，忠实为主。译文的第一稿在翻译过程中是最基本、最重要的。第一稿质量的高低，将影响翻译工作的进程，而顺利的进程能收到事半功倍的效果，必须认真对待。在进行这一道工序时，要着眼于译文的忠实、通顺，并以忠实为主。

（2）二稿核对，注意逻辑。俗话说："智者千虑，必有一失。"在翻译过程中，不出错误、一稿而成的人几乎不存在。即使初稿译得不错，也必须认认真真地进行校对，检查译文中的错漏之处，注重译文的前后逻辑关系，保持专业文体的逻辑性。

（3）三稿定局，润色词句。在前两稿的基础上，第三稿的重点是放在文字润色上。逐段、逐句、逐字细心删改、推敲，做到一忌含混，要文从字顺；二防文字累赘，求句炼词精。

1.2.2.3　翻译的方法

翻译的方法大体有直译和意译两种。所谓直译，是指既忠实于原文内容，又忠实于原文形式的翻译；所谓意译，是指只忠实于原文的内容，不拘泥于原文形式的翻译。直译和意译是达到同一翻译目标的两种翻译方法，两者相互补充、相辅相成。在进行翻译时，能直译的就直译，能意译的就意译，两者并用不悖。

2 专业英语翻译

英语是一种词性富于变化、词序比较灵活的拼音语言，而汉语则恰恰相反，是一种缺乏词序变化、词序比较固定的非拼音语言。由于一词多义的缘故，两种语言的单词又不能对应，单一词义的选择和引申还是不够的。因此，要在正确理解原文的基础上，完全摆脱原文表层结构的束缚，自由运用汉语来表达原文的意思，就必须应用各种翻译技巧，如词类转换、句子成分转换、词序和句序的调整等。

在英译汉的过程中，有时不一定机械地把英语某一词类同样译成汉语的同一词类。为了使译文通顺自然，常需进行词类转换，如把英语名词译为汉语动词，英语动词译为汉语名。

2.1 词性的转换

2.1.1 非动词译成动词

相比较而言，汉语中的动词用得较多，一个英语句子中只有一个谓语动词，而汉语句子中却可以有几个动词。以下举例说明。

(1) The control unite of a computer causes the machine to

operate according to man's wish.

译：计算机的控制单元使机器按照人的意志运转。

注意：句中除原来的谓语动词"causes"之外，不定式"to operate"和介词"according to"都按汉语习惯译成了动词。

（2）He admires the president's stated decision to fight for the job.

译：他对总统声明保住其职位而决心奋斗表示钦佩。

注意：句子中的分词"stated"、动词派生的名词"decision"、不定式"to fight for"都译成了动词，由此可以看出，汉语没有词性变化，但可以几个动词连用，因此英语中的不少词（名词介词、形容词和副词）在汉译时往往可以转译为动词。

2.1.1.1　名词转译成动词

动名词、动词派生的名词，具有动作意义的名词可以直接译为动词。以下举例说明。

（1）Control of dissolved oxygen（DO），solids retention time（SRT）and hydrolytic retention time（HRT）is necessary for efficient treatment of wastewater.

译：控制溶解氧（DO）、固体停留时间（SRT）和水力停留时间（HRT）对废水的有效处理非常重要。

（2）The main object of sedimentation is the separation of clear water from mixture.

译：沉淀的主要目的是将清水从混合液中分离出来。

一些加“er”或“or”的名词，有时在句中并不表示一个人的身份或职业，而具有较强的动作意义，这种词汉译时可译成动词。以下举例说明。

(1) Professor Wang was the instructor of our experiment.

译：王教授指导我们的实验。

注意：句中的“instructor”也不能译为“导师”，而应译为动词“指导”。

(2) Japan has become the major player in membrane market.

译：日本在膜市场上起重要作用。

注意：与上句相类似，“player”在这里也不能译成名词“选手”，而应将它译成动词“起……作用”。

适用于动词短语或介词短语中的名词可译为动词，如：pay attention to，make use of，with the help 等。

(1) These data has been made use of in production.

译：生产中已利用了这些数据。

(2) A body is negatively charged when it has electrons in excess of its normal number.

译：物体中所含电子超过正常数目时，物体带负电。

2.1.1.2 形容词译成动词

英语中表示感觉、知觉、信念的词，如：familiar，confident，sensible of 等，在句中作为表语时，译成动词。以下举例说明。

(1) Scientists are confident that all matter is indestructible.

译：科学家深信一切物质是不灭的。

（2）They are quite content with the data obtained from the experiment.

译：他们满足于实验中获得的数据。

（3）We are not sure about the effect of this parameter on the quality of the effluent.

译：我们不能确定这个参数对出水水质的影响。

有些要求有固定介词的形容词在句中作表语或定语时，译成动词。以下举例说明。

（1）The final product moisture is dependent on feed size and residence time at temperature.

译：最终产品的水分取决于入料粒度及干燥时间。

注意：句中将作表语的形容词短语“dependent on”译成动词“取决于”。

（2）LRCs are present in thick seams with less overburden than bituminous coals.

译：低阶煤的煤层厚，岩石覆盖比烟煤薄。

（3）For larger size feed，dried product moisture levels are higher.

译：如果入料粒度较大，干燥后产品的水分就较高。

注意：句中将作定语的“larger”和作表语的“higher”都译为动词。

起形容词作用的现在分词、过去分词和一些作定语或补语的形容词译成动词。以下举例说明。

（1） Figure 5 shows the rising power consumption at increasing sponge iron rates.

译：图 5 显示海绵铁配比增加时，耗电量增高。

注意：将此句中作定语的两个现在分词“rising”和“Increasing”译成动词“增加”和“增高”。

（2） While HWD is an additional processing cost，LRCWFS become a value added product capable of competing in energy markets.

译：尽管热水干燥增加了加工费用，但低阶煤水燃料已成为一种能够在能源市场上竞争的增值产品。

注意：将句中作定语的形容词“additional”译成动词“增加”，将作补语的形容词“capable”译成动词“能够”。

2.1.1.3 介词译成动词

某些具有动作意义的介词在汉译时可译成动词。以下举例说明。

（1） Numerous treatment technology of wastewater exist，each with their respective merits.

译：现在有很多水处理技术，每一种都具有各自的优点。

注意：句中将介词“with”译成汉语的动词“具有”。

（2） Conversely，for medium particles below limiting size，performance is relatively an affected and the geometry of the lower part of the cyclone becomes less important.

译：相反，如果介质的粒度低于限定的粒度，分选效果也就不太受影响，旋流器下部的几何形状也就不太重要了。

注意：句中将介词“below”译成动词“低于”。

(3) In a dense-medium cyclone with an unstable suspension as the medium, conditions in the lower part of the cyclone must be correct the raw coal particles are to be separated properly.

译：在利用不稳定悬浮液作分选介质的重介旋流器中，为了使原煤能够得到精确的分选，必须对旋流器下部的几何形状进行校正。

注意：句中将介词“with”译成动词“利用”。

2.1.1.4 副词译成动词

英语中的有些具有动作意义的副词，在和系动词 be 作联合谓语、宾补或状语时，可译成汉语的动词。以下举例说明。

(1) The oil in the tank is up.

译：油箱里的油用完了。

注意：句中将与“be”作联合谓语的副词“up”，译成动词“用完”。

(2) The experiment in chemistry was ten minutes behind.

译：化学实验耽误了十分钟。

注意：句中将作宾补的副词“behind”译成动词“耽误”。

（3） The reaction force to this action force pushes the rocket ship along.

译：这个作用力的反作用力推动宇宙飞船前进。

注意：与上句相同，将作宾补的副词“along”译成动词“前进”。

2.1.2 非名词译成名词

2.1.2.1 动词译成名词

英语中有很多由名词派生的动词以及由名词转用的动词，汉译时往往找不到相应的动词，可将其译成汉语的名词。以下举例说明。

（1） Coating thickness range from one-tenth mm to 2 mm.

译：涂层厚度的范围为0.1~2毫米。

注意：句中将名词转用的动词“range”译为名词“范围”。

（2） An electron or an atom behaves in some ways as though it were a group of waves.

译：电子或原子的运行方式，多少有些像一组波动。

注意：句中把名词转用的动词“behaves”译成名词“运行方式”。

2.1.2.2 形容词译成名词

除了“the+形容词”表示一类人或物之外，表示事物特征的形容词作表语时，往往在其加“性、体、度”等，译成名词。以下举例说明。

（1）Computers are more flexible and can do a greater variety of jobs.

译：计算机的灵活性较大，因此能做更多不同的工作。

注意：句中将表示计算机特征的形容词“flexible”译成名词“灵活性”。

（2）This steam engine is only about 20% efficient.

译：这台蒸汽机的效率只有20%。

注意：句中将表示“steam engine”特征的形容词“efficient”译成名词“效率”。

（3）Glass is much more soluble than quartz.

译：玻璃的可溶性比石英大得多。

注意：句中将表示“glass”特征的形容词“soluble”译为名词“可溶性”。

（4）Mercury is appreciably volatile even at room temperature.

译：即使在室温下水银的挥发性也很显著。

注意：句中表示“mercury”特征的形容词“volatile”译成名词“挥发性”。

2.1.2.3 副词译成名词

某些由名词派生的副词在表示“用……方法、在……方面”

等意义时，可译成汉语的名词。以下举例说明。

（1）The equipment employed in the commercial test is shown schematically in Figure 5.

译：本次工业试验中所用设备的简图如图 5 所示。

注意：句中“schematically”的原意是“用简图的方法”，在此将它译成名词“简图”。

（2）Oxygen is one of the important elements in the world，it is very active chemically.

译：氧是世界上的重要元素之一，它的化学性质很活泼。

注意：句中“chemically”原意是“在化学性质方面”，在此将它译成名词“化学性质”。

（3）Gold is an important metal but it is not essentially changed by man's treatment of it.

译：金是一种很重要的金属，但人为的加工不能改变它的性质。

注意：“essentially”的原意是“本质上”，在此将它译成名词“性质”。

2.1.3 非形容词译成形容词

2.1.3.1 名词译成形容词

用“be+of+名词”表达事物性质的名词，常译成形容词。以下举例说明。

Robots are now in use in industrial plants throughout the world, they are proved to be of great ability.

译：今天机器人在世界各地的工厂中已得到广泛应用，并已证明是十分能干的。

注意：句中“be of great ability”相当于“very able”。

有些名词加不定冠词 a 或 an 作表语时，译为形容词。以下举例说明。

（1） Their experiment is a success.

译：他们的实验很成功。

（2） As he is a perfect stranger in the city, I hope you'll give him the necessary help.

译：由于他对这个城市完全陌生，我希望你能给他必要的帮助。

（3） Independent thinking is an absolute necessity in studying.

译：学习中独立思考是完全必要的。

由形容词派生的名词往往可译成形容词。以下举例说明。

（1） He found there were many difficulties to design this complicated sewage plant without a computer.

译：他感到没有一台计算机要完成这一复杂的污水厂设计是很困难的。

注意：句中把由名词“difficulty”派生的名词“difficulties”译成形容词“困难的”。

（2） The complication of mathematical problems made him into difficulties complicated.

译：复杂的数学问题使他陷入困境。

注意：句中派生的名词“complication”译成形容词“复杂的”。

（3） The combination of mechanical properties of this alloy can be well achieved by heat treatment.

译：通过热处理能够获得这种合金的综合机械性能。

注意：句中把由形容词“combinative”派生的名词“combination”译成形容词“综合的”。

2.1.3.2 副词译成形容词

当动词译成名词或形容词译成名词后，原来修饰动词或形容词的副词就转译成形容词。以下举例说明。

This communication system is chiefly characterized by its simplicity of operation and the ease with which it can be maintained.

译：这种操作系统的主要特点是操作简单，易于维修。

当副词作定语时译成形容词，这种副词多放于被修饰词之后。以下举例说明。

（1） The table below shows the specific gravities of metals.

译：下表给出了各种金属的密度。

（2） The power stations here supply the whole city electricity.

译：这里的发电厂供给全市的电力。

2.2　被动语态的翻译

科技英语的一大特点就是在文中大量地使用被动语态。与英语相比，汉语中虽也使用被动语态，但远不如英语那样广泛。往往英语中用被动语态表达的句子，汉语中却用主动语态表达。如定义“能”时，英语中用“Energy is defined as the ability to do work”这一被动句来表示，而汉语中却用主动句“能的定义为做功的能力”来表达。因此，在翻译被动语态时，必须摆脱原句的语态限制，凡能译成主动句的，应尽量译成主动句，不宜译成主动句的才译成被动句，这样翻译才符合汉语的表达习惯。

2.2.1　译成汉语的主动句

2.2.1.1　原文中的主语仍作主语

当原文中的主语为无生命的名词，且句中没有由 by 引导的行为主体时，原文中的主语仍作主语。以下举例说明。

（1）Air-sampling devices are used to detect and measure smoke, particles, and gases.

译：大气采样器用于监测和测量空气中的烟尘、颗粒物和气溶胶。

（2）Simply stated, the environment can be defined as one's surroundings.

译：简单来说，环境的定义为我们周围的环境。

（3）Our culture as well as our aesthetic heritage is also being lost to pollution.

译：我们的文化遗产和艺术遗产正在遭受污染。

（4） Many believe that these changes are caused by acidic deposition traceable to pollutant acid precursors that result from the burning of fossil fuels.

译：许多人相信是酸沉降导致了这些变化，而这些酸性污染物则是源于化石燃料的燃烧。

2.2.1.2 原文中的主语在译文中作宾语

当被动句中有地点状语或有介词 by、from、at、in 等引导的状语时，将此状语译成主语，原主语译为宾语。以下举例说明。

（1） The concentration of secondary pollutants will be influenced by the same gross atmospheric features, such as air mass origin and degree of dispersion, as the primary pollutants.

译：与一次污染物相同，像气团源和分散度等大气特性会影响二次污染物的浓度。

（2） Living spores of various fungi have been collected by an aeroplane above the Caribbean Sea, 800 miles from their nearest source.

译：在加勒比海上空，距最近的污染源 800 英里的飞机上可检测到不同真菌的活性孢子。

（3） In spite of the usually low concentration at the source, marine bacteria have been collected 80 miles inland from the nearest coast.

译：尽管污染源的浓度通常很低，但在离最近海岸 80 英里的岛上已收集到了海洋细菌。

某些要求宾语或宾补的动词用于被动语态，翻译时需要在其前加“人们、大家、有人”等具有广泛意义的词作主语，原来的主语译成宾语。以下举例说明。

（1） He was seen to be working in the workshop.

译：有人看见他在车间工作。

（2） Silver is known to be the best conductor.

译：我们知道，白银是最好的导体。

（3） Rubber is found a good insulating material.

译：人们发现橡胶是一种良好的绝缘材料。

在这三个短句中，谓语动词“seen，known，found”都既带有宾语“him，silver，rubber”，又分别带有宾补“to be working in the workshop”“to be the best conductor”和“to be a good insulting material”。因此，翻译时要增加主语。

2.2.1.3 译成用“的”字结构做主语或表语的判断句

用“的”作表语的判断句可用来翻译某些被动句，用“的”作主语的判断句尤其适合于用来翻译倒装语序的被动句。

译成用“的”字结构作表语的判断句，以下举例说明。

（1） Rainbows are formed when sunlight pass through small drops of water in the sky.

译：彩虹是当光线穿过天空中的小水滴时形成的。

（2） These analytical equipment are imported from US.

译：这些分析仪器是从美国进口的。

（3） These reports should be fully utilized by us when studying a project，as they are the most authoritative records available.

译：这些报告是我们在研究项目时应该充分利用的，因为它们是可利用的最有权威的资料。

译成用“的”字结构作主语的判断句，以下举例说明。

Produced by electrons are the X-rays, which allow the doctor to look aside a patents.

译：电子产生的是 X 射线，医生用它做透视。

2.2.1.4 常用被动句型的翻译

有一类以“it”作形式主语的被动句，翻译时常常需要译成主动形式，有时可不加主语，但有时却要加上不确定的主语，比如“有人、大家、人们、我们”等。

不加主语的，举例如下。

It is hoped that… 希望……

It is reported that… 据报道……

It is said that… 据说……

It must be supposed that… 据猜测……

It must be admitted that… 必须承认……

It must be pointed out that… 必须指出……

It will be seen from this that… 由此可以看出……

加主语的，举例如下。

It is believed that… 有人相信……

It is generally considered that… 大家认为……

It is well known that… 大家知道……

It was told that… 有人说……

It is asserted that… 有人主张……

It is found that…　　　　人们发现……

It is stressed that…　　　　人们强调……

It has been suggested that…　　　　有人认为……

2.2.2　译成汉语的被动句

汉语有时也采用被动表达，这一类句子都强调被动的动作，有的说出了动作的主动者，有的则不说出动作的主动者。英语的被动句译成汉语的被动句往往采用“被、给、由、受、为……所”等表示被动的字眼。以下举例说明。

（1）The sewage of nearly 10 million people in the US is discharged raw into our waterways.

译：美国近一千万人口的生活污水未经处理直接被排入我们的下水道。

（2）When these compounds decomposed by bacteria，oxygen is removed from water.

译：当这些组分由微生物降解时，氧就被从水中排除。

（3）It has been suggested that plants be used as indicators of harmful contaminants because of their greater sensitivity to certain specific contaminant armful contaminant.

译：有人认为植物可被用作有害污染物的指示剂，因为它们对某些特定的污染物质非常敏感。

（4）In this study，those remaining effects are treated as stochastic noise and are assumed white gaussian distributed with zero mean.

译：在这项研究中，那些未被考虑的各影响因素被作为随时噪声加以处理，并假设其具有零均值白高斯分布。

2.2.3 译成汉语的无主句

科技文章在讲述事情时，常强调怎么做而不介意谁去做，这样许多被动句可译成无主句。翻译时，原被动句中的主语就成了无主句的宾语；有时还可以在原主语之前加“把、将、使、对、由”等。以下举例说明。

（1） After reducing the BOD to nominally 60mg/L, wastewater treated can be directly discharged into nature water body.

译：将废水中的BOD降到60mg/L以后，处理后的水可直接排入天然水体。

（2） Solid particles must be removed from gaseous effluents because, if they are not removed, they settle out on land and houses, or in people's lungs.

译：必须将固体颗粒从气态排放物中去除，否则它们会沉降在土壤、房屋上或沉积在人的胃里。

（3） The solids wastes found in the petrochemical industry may be stored, handled, and disposed of by various combinations of many different methods.

译：可对石油化学工业产生的固体废物进行储存、处理，并可采用不同的方法，通过不同的结合方式进行处置。

（4） Usually, before the municipal solid waste can be landfilled, it does require digestion to avoid odors insects and water pollution.

译：通常在把城市固体废物进行卫生填埋之前，必须进行消化处理，以防散发臭味，滋生昆虫，污染水体。

2.3　定语从句的译法

从和主句关系的密切程度来说，定语从句可分为限定性定语从句和非限定性定语从句。限定性定语从句和主句的关系比较密切，而非限定性定语从句和主句的关系较为松散。正是由于在文中广泛地使用定语从句，才使得科技文章的句子结构和句子之间的逻辑关系比较复杂。

2.3.1　前置法当定语

从句不太长或不太复杂时，可将其译成“的”字结构放在被修饰词的前面。以下举例说明。

（1）It is the function of the medium recovery system to recovery the magnetite that is rinsed from the products on the rinse screen and to remove the nonmagnetic materials from a portion of the main medium circulation system for viscosity control.

译：介质回收系统的作用是回收经喷淋脱介筛上冲下的磁铁矿和除去部分主要介质循环系统中的非磁性物以控制黏度。

注意：句中“function of the medium recovery system”的主语由“to recovery the magnetite”和“to remove the nonmagnetic materials”两部分组成。因此，可将这个长句分成两个分句“to recovery the magnetite that is rinsed from the products on the rinse screen", "to remove the nonmagnetic materials from a portion of the main medium circulation system for viscosity control"。在第一个分句中,"that is rinsed from the products on the rinse screen"是定语从句,修饰"magnetite";在第二个分句中,"for viscosity control"作目的状语。

(2) The earth could be cooled, conceivably to the point of initiating another ice age, if atmospheric contaminants sufficiently reduce the amount of solar energy that penetrates the earth's atmosphere.

译:如果大气中的污染物能够有效地减少穿过大气层的太阳能,地球将被冷却,到一定程度也许会出现另一个冰河时代。

(3) We are a long way to go from a technology that will restore the particulate matter level of the atmosphere to that of a century or more ago.

译:要发明一项技术,使大气中颗粒物恢复到约一个世纪以前的水平,我们还需经过长期的努力。

2.3.2 分译法

把定语从句译成主句的并列句,放于主句之后。分译法适用于以下两种情况:

(1) 限定性定语从句较长;

(2) 限定性定语从句虽不长,但先行词的修饰成分较多。

如果采用前置法将定语从句放于被修饰词之前,就会使译文层次模糊,表达不清。分译法有两种形式:重复先行词和省略先行词。

2.3.2.1 重复先行词

重复先行词或用“他、它、他们、它们、这些、那些”等代词来代表先行词。以下举例说明。

（1）Corrosion is an electro－chemical process by which metal, such as, mild steel, returns to its natural state, such as iron oxide or rust.

译：腐蚀是一种电化学过程，在这一过程中像低碳钢之类的金属会恢复到其自然状态，如铁以氧化铁或铁锈状态存在。

注意：句中由“which”引导的定语从句修饰“process”，而“process”还有前置定语“an electrochemical”，并且从句在语义上是为了进一步说明，故为了使译文层次分明，表达清晰，采用分译法，并重复先行词。

（2）There is an ozone layer in the lower stratosphere that absorbs an appreciable part of the energy in the ultraviolet wavelengths emitted by the sun.

译：臭氧层存在于低层平流层中，它能吸收太阳辐射的紫外线长波的部分能量。

（3）The contaminants whose buildup in the atmosphere could hypothetically cause reduced penetration of solar energy are suspended particulates that have the ability to absorb and scatter solar energy.

译：在大气层中，假定能减少太阳能透射的污染物是一些悬浮颗粒，这些颗粒能吸收和分散太阳能。

注意：句中有两个定语从句，分别为修饰“contaminants”的从句“whose buildup in…penetration of solar energy”和修饰“suspended particulates”的从句“that have the ability…solar energy”。翻译时考虑到译文结构紧凑，逻辑性强，第一个从句采用前置法，第二个则采用分译法。

（4） The gases pollutants are usually formed in homogeneous gas-phase reactions that in many cases are photochemical initiated.

译：气态污染物通常是在均质的气相反应中形成的，多数情况下，这些反应都是由光化作用推动的。

注意：句中“which”引导的定语从句虽不长，但其先行词“reactions”的修饰成分比较多，需采用分译法，重复先行词。

（5） Dust and other particles matter in the air provide nuclei around which condensation takes place, forming droplet and thereby playing a role in snowfall patterns.

译：尘粒和空气中的其他颗粒物提供了核，在这些核周围发生浓集作用形成了微小颗粒，进而在降雨或降雪中起到一定的作用。

2.3.2.2 省略先行词

有时省略先行词代表的意义也能达到层次分明、语义清楚的目的。以下举例说明。

（1） The activated silica gel can be composted into flocculants which are being tested in domestic sewage to remove nitrogen and phosphorous.

注意：将主句和从句都译成主动句；如果在从句中仍重复先行词，会使句子显得很啰嗦。

未省略先行词的翻译为：活化后的水玻璃可制成复合型的絮凝剂，复合型絮凝剂用于生活污水脱氮除磷的试验研究。

省略先行词的翻译为：活化后的水玻璃可制成复合型的絮凝剂，用于生活污水脱氮除磷的试验研究。

（2）Devolatilized tar，being hydrophobic，remains on the coal surface in the pressured aqueous environment，producing a uniform coating that seals the microporous limits moisture reapportion.

这个句子的逻辑关系比较复杂：

1）“being hydrophobic”可以有两种理解，对主语“Devolatilized tar”进行补充说明或作为“remains on the coal surface”的原因。

2）分词短语“producing a uniform coating”是“remains on the coal surface”的结果。

3）“seals the microporous limits moisture reapportion”是“a uniform coating”的结果。

4）连词“and”连接的两个动词“seals”和“limits”也有逻辑关系，“limits moisture reapportion”是“seals the microporous”的结果。

注意：在一般句子中，要补译出表示句子之间逻辑关系的关联词，但此句的逻辑关系比较复杂，如果将这些关联词补译出来，会使句子显得十分啰唆。

加关联词的翻译为：经脱挥发分作用得到的焦油由于是疏水的，因此在高压的液相中仍然留在煤的表面，形成一层均匀的油膜，从而封住了煤中的微孔，所以减少了煤对水分的重新吸附。

不加关联词的翻译为：经脱挥发分作用得到的焦油是疏水的，在高压的液相中仍然留在煤的表面，形成一层均匀的油膜，封住了煤中的微孔，从而减少了煤对水分的重新吸附。

（3） Figure 1 incorporates many factors, which must be considered in developing a satisfactory stem.

译：图 1 所示的许多因素，在研制性能良好的系统时必须予以考虑。

（4） The energy obtained from uranium atoms in nuclear-power stations may be used to heat boilers and produce steam for the turbines that derive the alternators.

译：从原子能发电厂的铀元素中获得的能量可以用来加热锅炉、产生蒸汽、驱动汽轮机，以带动交流发电机。

2.3.3 融合法

把原句中的主语和定语从句融合在一起，用原句的主语作主语部分，原句中的定语从句作谓语部分，译成一个独立句子。

“There be” 结构中的定语从句在翻译时常采用融合法，举例如下。

（1） There is some metal， which is lighter than water.

译：有些金属比水轻。

（2） There are many new technologies and devices which are used to remove the nitrogen and phosphorous in wastewater.

译：很多新技术和新设备被用于去除废水中的氮和磷。

注意：将“there be”句中的“many new technologies and devices”译成主语，定语从句中除“which”以外的部分作谓语，同时将它译成被动语态。

（3） There are many diseases， which are associated with the contamination of water supplies by animal or human wastes.

译：很多种疾病与因人畜排泄物所引起的供水污染有关。

除了“there be”结构中的定语从句外，其他一些定语从句也可采用这种方法进行翻译。以下举例说明。

（1） The chemical properties of a certain element that depend on the arrangement of these electrons， particular the outer， or valence electrons， are the same.

注意：如果将“which”引导的定语从句译成主句的并列句，会使句子的意思显得松散、不连贯。

分译：各元素的化学性质是一样的，这一性质取决于电子（尤其是外层电子或价电子）的排列方式。

合译：元素的化学性质取决于电子（尤其是外层电子）的排列方式，在这一点上各元素是一样的。

（2） Low mining cost，high reactivity，and extremely low sulfur content make these coal premium fuels if not for their high moisture levels which range from 25% to more than 60%.

分译：低开采成本、高反应活性以及很低的硫分会使这些 LRCs 成为优质燃料，如果不是由于它们的水分含量很高，这些煤的水分含量达 25%～60%。

合译：如果不是由于它们的水分含量高达 25%～60%，低开采成本、高反应活性以及很低的硫分会使这些 LRCs 成为优质燃料。

（3） Besides carbon dioxide，scientists identify other gases that are thought to contribute to global warming.

译：科学家已证实除了二氧化碳外还有其他一些气体也在促使全球变暖。

(2) Low mining cost, high reactivity, and [illegible] low sulfur content make these coal formation [illegible] if not for their high moisture levels which [illegible] 25% to more than 60%.

译文：低开采成本，[illegible]硫的含量使这些LRCs成为优质的燃料，如果不是由于它们的水分含量很高，这些煤的水分含量达25%～60%。

合译：如果不是由于它们的水分含量高达25%～60%，低开采成本、高反应活性以及很低的硫含量会使这些LRCs成为优质燃料。

(3) Besides carbon dioxide, scientists identify other gases that [illegible] to global warming.

译：科学家已经发现除了二氧化碳外还有其他一些气体也在促使全球变暖。

下 篇

英语实例

YINGYU SHILI

3 环境工程专业英语实例

3.1 水污染及控制技术

3.1.1 水污染及污染物

Water Pollution and Pollutants

In the recent decades, studies on waste water characteristics have drawn attention towards the environmental occurrence of a variety of newly identified compounds of anthropogenic origin. The occurrence of such trace compounds (mostly organic), known as the "emerging pollutants" and their harmful impact on both aquatic and terrestrial life forms as well as on human health is now an issue of concern among the scientists, engineers, and the general public as well. The non-regulated organic trace pollutants, known as emerging micro-pollutants, have been recently introduced or newly detected with the help of advanced analytical technologies. A contaminant whose new origin, alternate route to humans or new treatment techniques has been innovated is termed as "emerging". They are categorized by apprehensible, probable or actual risk to human health and environment. They may be industrial in origin or may originate from municipal (domestic), agricultural, hospital or laboratory wastewater. In large part, the compounds in question are derived from three broad categories, viz.

(1) Pharmaceuticals (PhACs),

(2) Personal Care Products (PCPs),

(3) Endocrine Disrupting Compounds (EDCs).

But they are not confined to the above and may comprise of nanomaterials (NMs), metabolites of Emerging contaminants (ECs), illegal drugs, engineered genes, etc. NMs affect the bacterial biomass during waste water treatment and thereby decrease their biological activity leading to decrease in EC removal efficiency. ECs are present and have been found in surface water, ground water as well as drinking water and in wastewater treatment plants (WWTP) discharge.

Municipal wastewater is viewed as one of the principle discharge sources for the emanation of emerging contaminants like non-point and point sources, industries and stormwater, wastewater from households and water treatment facilities into the environment. Also there is a growing concern of sludge management due to high levels of ECs in them. The design of the current WWTP could not restrict the elimination of the emerging contaminants and their metabolites where it is released into rivers or streams having high biodiversity as sewage effluents. So far, considerable work have been done in regard to the performance of wastewater technologies in case of nutrient removal, while there is an absence of data on the ability to removal of ECs, and additionally on the adverse ecotoxicological impacts of these compounds on surface water bodies.

Pharmaceutical molecules identified in the wastewater belong to several classes of human and veterinary antibiotics, human prescription and non-prescription drugs, and some sex and steroid hormones as well. Personal care products (PCPs) include chemicals found in consumer products (e. g. galaxolide, tonalide). Endocrine disrupting compounds (EDCs) can elicit adverse effects on endocrine systems as they have androgenic or estrogenic activities even at low concentrations, as shown in Fig. 3. 1.

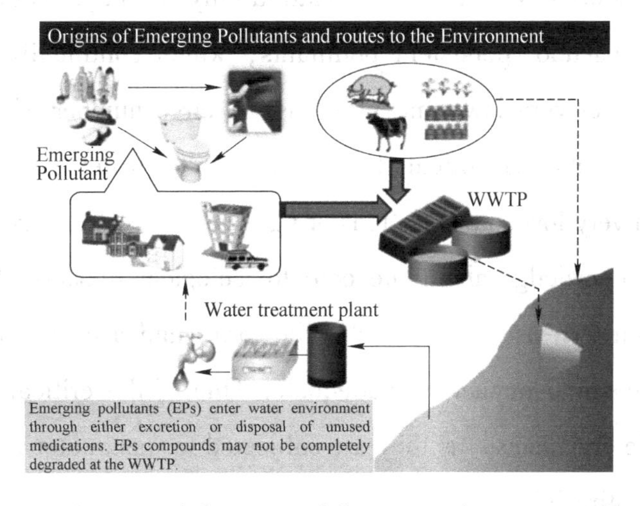

Fig. 3. 1 Conceptual depiction of the origin of emerging pollutants

Pharmaceuticals (PhACs)

PhACs are a set of developing ecological contaminants that are broadly and progressively being utilized as a part of human and veterinary medication. They include compounds of environmental concern like antibiotics, legal and illicit drugs, analgesics, steroids, etc. Their persistence in the body occurs due to their specific mode of action. They have been detected in WWTP effluents, sludge, sediments, natural waters, drinking water and groundwater. They are supposed to provoke the development of antibiotic resistant genes in soil bacteria. Nowadays, active pharmaceutical ngredients (APIs) and their biotransformation products, which are largely unstudied, are bioaccumulating and causing significant consequences to ecosystem (Arnold et al., 2013). Although these compounds have been entering the environment for many years, but investigation on their adverse effects on aquatic organisms have started only recently. They are considered as pseudo-persistent pollutants, which continually enter the environment at very low concentrations. A large number of more than 160 different pharmaceuticals have as of now been detected in aquatic systems in very low concentrations of ng/L to low μg/L range. There is very little knowledge about the eco-toxicological impacts of pharmaceuticals on terrestrial and aquatic life forms and a complete analysis eco-toxicological impact is inadequate. One of the critical targets is the aquatic organisms, as they are subjected to wastewater remnants over their entire life.

Personal care products (PCPs)

PCPs are yet another class of emerging contaminants that incorporate prescribed and non prescribed veterinary and human pharmaceuticals and the agile and inert elements for individual care purposes. A few PCPs to name are cosmetic products, engineered hormones, steroids, perfumes, shampoos, etc. UV filters, known to have estrogenic activity, are reported to be one of the most commonly found PCPs in ground water and other aquatic environment. PCPs are released into wastewater and advance toward WWTPs, in their native or biologically transformed structures. In WWTPs, the likely fates of PCPs and their metabolites are conversion to CO_2 and water; mixing with the receiving water bodies either as the original or mineralized product; and sorption by the solids like sludge/biosolids, mainly if the compound or the biologically moderated transformation product is lipophilic.

Endocrine disruptors (EDCs)

EDCs are characterized as the artificial chemicals that, when ingested into the body can either copy or obstruct hormones and effect body's normal functioning. The Environmental Protection Agency (EPA) characterizes EDCs as external agents that meddle with the formation, release, transport, attachment, activity, or displacement of body's natural hormones that maintains homeostasis, development, reproduction and behavior (United States Environmental Protection Agency, USEPA, 1997). It is by and large acknowledged that the three main classes of EDCs are estrogenic i. e. they mimic or alters the functioning of natural estrogens, androgenic (copy or obstruct natural testosterone) and thyroidal (causes immediate or oblique effects to the thyroid). Natural and engineered EDCs are discharged into the environment by human activities, creatures and industries; essentially through sewage treatment systems before finally going to soil, surface water, silt and ground water. Preponderance research has concentrated just on estrogenic compounds. EDCs are present in immensely low concentrations (ng/L or μg/L) in wastewater. These compounds are of profound concern as their long term exposure and adverse impact on human are unknown.

Vocabulary

anthropogenic [ænθrəpəˈdʒenɪk] *a.* 人为的
emerging contaminants 新兴污染物
pharmaceuticals [ˌfaːməˈsjuːtɪklz] *n.* 药物
personal care products 个人护理品
endocrine disruptors 内分泌干扰素
estrogens [estrədʒəns] *n.* 雌激素

Reading materials A

Environmental/Health Issues and Regulations Related to Emerging Contaminants

Due to the absence of relevant data on the impacts, fate and concentration levels of emerging contaminants make it troublesome for overnments to control their utilization and also manage the levels that re already persisting in the environment. There are as of now no laws mandates illustrating the upper limits of concentrations of emerging ontaminants in wastewater discharge, drinking water, or the environment. In United States, a preparatory observing technique was rganized and an archive was communicated illustrating the preparatory way to deal with EDCs and to decrease its intrusion in people nd wildlife (European Commission, 2011). This document concentrates on the diminishment of the utilization of EDCs in consumer roducts, sustenance added substances, and beauty care products, however does not set any proposals for uttermost permissible concentrations in drinking water, wastewater and in nature. The European Union Water Framework Directive isted 45 priority compounds with environmental quality tandard (EQS) to be respected in aquatic environments and listed 10 thers on contemporary watch list (Decision 2015/495, published on 4th march 2015). Similar regulations were followed by Switzerland for several ECs. In 1995, the European Union (EU) set 10ng/L and 10μg/kg as the concentration of PhACs and PCPs in surface water and soil. The U. S. Food and Drug Administration (FDA, USA) publicized directions for the evaluation of human drugs. Environmental assessment has reported the expected introduction concentration of pharmaceuticals in the aquatic environment as ≥1μg/L (US FDA, 1998). EAWAG Institute, Switzerland proposed similar to EQS environmental quality criteria for several

ECs such as pharmaceuticals and hormones (PPHs) and pesticides. Many disinfection byproducts that are transformation products of ECs after treatment are regulated by the US, European Union and World Health Organization (WHO). But in Canada and India, there is no such regulation for ECs.

The Water Framework Directive included anti-inflammatory diclofenac or the synthetic hormones Ethynylestradiol (EE2) in the supposed "watch list" of priority compounds to address the risk posed by these substances. Various PhACs and EDCs, were enlisted in the Drinking Water Contaminant Candidate List (US EPA, 2012). Different PhACs, for example, carbamazepine, naproxen, sulfamethoxazole, ibuprofen, gemfibrozil, atenolol, diclofenac, erythromycin and bezafibrate have been rated prime concern pharmaceuticals to the water cycle by the Global Water Research Coalition.

Contaminants from PhACs, PCPs and EDCs enter water bodies and can exceed and persist beyond acceptable levels. The widespread occurrence of ECs in water has high probability of their incorporation in crops irrigated with contaminated water and posses risk to human health upon consumption. ECs can cause harmful impacts on aquatic and terrestrial wildlife and human communities. Endocrine disrupting chemicals cause a number of reproductive and sexual abnormalities in wildlife and humans. Subjection to these chemicals amid pre and postnatal life, can impair the development and signaling of the endocrine system. The effects during development are permanent and sometimes irreversible. Managing ECs in water resources is a critical issue that requires attention especially in sensitive ecosystems and in rapidly developing areas. However, the ecological effects of ECs in natural environment are different from laboratory tests. When present in the environment many factors like pH, soil or water type, ionizable compounds, etc. can influence th bioavailability of the ECs. There is a need for a comprehensive framework that aims at system - wide abatement (source transfer–fate levels) using both structural and non–structural approaches.

Reading materials B

Emerging Contaminants in Wastewater

Municipal wastewater treatment plants (WWTPs) are by and large not furnished to manage complex pharmaceuticals, as they were constructed and updated chiefly with the intension to eliminate effortlessly or modestly biodegradable carbon, phosphorus and nitrogenous substances and microbes, which consistently appear at the WWTPs in μg/L levels. Parent human pharmaceuticals or their metabolites enter the aquatic systems through WWTPs and PCPs (e.g. perfumes) are released through shower waste. Countries like USA, Japan, United Kingdom, Finland and Spain had documented the presence of PhACs and PCPs in concentrations of ng/L to μg/L in WWTPs. Nowadays engineered nanomaterials from consumer products are entering the environment which is seldom detected but can have inimical affects. The primary origin of steroidal hormones in aquatic environment is human and animal defecation. In the long run the natural and engineered hormones and their metabolites finally reach WWTP.

Hormone replacement therapy (HRTh) along with oral dose of progesterone, estrogens and at times testosterone, can add to the absolute estrogenicity of municipal wastewater. The conceivable removal course of the hormones from various treatment techniques can be categorized under four procedures, viz. abiotic degradation, biological degeneration, volatilization and adsorption onto solids.

The traditional wastewater treatment systems for the most part comprises of an primary treatment, secondary treatment and occasionally a tertiary step, with various biological and physicochemical procedures accessible for every phase of the treatment. In the primary treatment the solid waste substances of the wastewater such as settleable solids, plastics, oils and fats, sand and grit, etc., are separated. This method is common for almost all Urban wastewater treatment plant (UWTP) and is accomplished mechanically by filtration and sedimentation. Nonetheless, the secondary treatment, which normally depends on the biological (aerobic or anaerobic) degradation of organic substances or nutrients, can vary significantly. Among the various biological treatment techniques e.g. fixed bed bioreactors (FBR), Membrane bioreactors (MBR), moving bed biofilm reactor (MBBR) etc., used in UWTPs, the most well-known technique is conventional activated sludge (CAS). Organic substances and nitrogen are removed under certain conditions by activated sludge plants through the formation of biological floc utilizing dissolved oxygen. Lastly in the tertiary treatment phosphorous can be removed by precipitation and filtration. Also some UWTPs disinfection of the effluent is done by UV irradiation or chlorination, before discharging them in the environment. But these treatments alone do not ensure complete removal of the ECs. The most common activated sludge technique which is used worldwide cannot remove all PPCPs efficiently and entirely e. g. diclofenac and carbamazepine that are resistant to biodegradation. Moreover, various processes like biological and chemical degradation and photolysis may transform ECs into forms that can be more toxic than their parent compound. Partial oxidation of PhACs leads to generation of transformation products (TPs) with more ecotoxicity such as N-(4-carbamoyl-2-imino-5-oxoimidazolidin)-formamido-N-methoxyacetic acid (COFA) and carboxy-acyclovir are the two TPs of the antiviral drug acyclovir that are more harmful than cyclovir. Disinfection byproducts (DBPs) are yet another type of TP that is formed when organic matter present in water reacts with disinfection agents like ozone, chlorine, etc. Today more than 600 DBPs are reported and the number is ever rising.

摘自：Anindita Gogoi, Payal Mazumder, Vinay Kumar Tyagi. Groundwater for Sustainable Development 6 (2018) 169-180.

3.1.2　内陆水域与海洋污染

Pollution of Inland Waters and Oceans

The coast is a zone or strip of land extending from the coastline, which borders the sea to where the land rises inland. Its limit is marked by the level of high tide. The coastline is the triple interface of air, land and sea. The word pollution comes from the verb pollute, which means to make impure. The definition of coastal pollution by the World Health Organization goes like this "The introduction by man, directly or indirectly, of substances or energy into the marine environment, including estuaries, which results or is likely to result in such deleterious effects such as harm to living resources and marine life, hazards to human health, hindrance to marine activities, including fishing and other legitimate uses of the sea, impairment of quality for use of sea water and reduction of amenities."

The chemical and biological characteristics of coastal waters are really susceptible to addition of biodegradable and stable compounds from soil. Qasim and Sen Gupta predicted that in 1984, 5 million tonnes of fertilizers, 55000 tonnes of pesticides, and 125000 tonnes of synthetic detergents were used in India. On an average, 25% of all these can be anticipated to end up in the ocean every year. More or less of these substances are biodegradable while others are relentless. Their cumulative impact to the coastal marine environment, over a long period could be quite harmful.

It is mainly the human activities that are responsible for a major decline of the world's biological diversity, and the problem becomes more complicated when combined human impacts could have enhanced present loss rates to 1000 ~ 10000 times the usual rate. In the seas, the marine life faces threats in many ways, such as over exploitation and harvesting, deposit of waste, contamination, exotic species, soil recovery, dredging and global climate change. One of the major kinds of human impact constitutes a major threat to marine life: the pollution by plastic debris. So, the issue of pollution may be served to analyze the natural events in future. Anticipating the effects of contamination on marine population by the purpose of simulation models requires that all sources of pollution over the total orbit of the species be considered. Even though a particular pollutant, waste disposal site, or habitat loss may be judged minor when judged independently, the cumulative effect might be significant.

Pollution in marine coastal areas is also considered from point and non-point land-based sources, such as rivers, drainage ditches, submarine outfalls and coastal cities. The portion of pollutants in coastal regions is determined by the combination of three mechanisms:

(1) Advection by currents,

(2) Turbulent diffusion,

(3) Chemical, biological or other interactions.

In relatively shallow and coastal areas with weak tides, wind is the main mechanism for generation of currents. The shear stresses applied at the sea surface creates a three dimensional circulation, greatly influenced by the natural process of the Coriolis and inertia forces.

Causes

There can be many causes of coastal pollution. Depending upon the position, the extent of pollution varies. The primary origin of pollution being the humans can be separated as the pollution by humans on the soil and off the ground.

(1) Plastic debris

The only substance that is illegal to dump anywhere in the ocean is plastic. Plastics are primarily synthetic organic polymers derived from petroleum. Plastic materials are found to be the major macroscopic pollutants in many published reports about plastic debris found throughout the world. The versatility of these materials has led to a large increment in their use over the past three decades, and they have rapidly moved into all facets of daily life. Plastics are light-weight, strong, durable and cheap, characteristics that make them suitable for the manufacture of a very wide range of products. These same properties happen to be the reasons why plastics are a grave risk to the surroundings . Since they are also floating, an increasing load of plastic debris is being spread over long distances, and when they eventually settle in sediments they may be kept for centuries.

The polymers when exposed to UV radiation in sunlight break into smaller and smaller pieces, but they are still present as plastic, and they are non-biodegradable in any practical manner. This persistence of plastic leads to an increasing abundance in the ocean environment, which makes plastic debris more accessible to plankton and other marine life. Though these synthetic organic polymers have only existed for just over a century, from 1988 in the United States alone, 30 million tons of plastic were produced annually. Also it was found that plastic degradation process is slower in the ocean than land because ocean water which is cool hinders the reaction.

The threat of plastics to the maritime environment has been neglected for a long time, and its seriousness has been just recently recogmized. In the maritime surroundings, the perceived abundance of marine life and the immensity of the oceans have passed to the firing of the proliferation of plastic debris as a potential hazard.

Studies by Gregory and Ryan show that plastics are the predominant amongst the marine litter, and its proportion consistently varies between 60% and 80% of the total marine debris. On that point is however comparatively little data on the impact of plastics pollution on the ocean's ecosystems. A study done on 1033 birds collected off the coast of North Carolina in the USA found that individuals from 55% of the species recorded had plastic particles in their guts. Minute particles floating in the ocean are consumed by birds that resemble their natural food. Over the past 20 years polychlorinated biphenyls (PCBs) have increasingly polluted marine food webs, and are prevalent in seabirds.

(2) Sewage and effluents

Although it is hard to imagine raw sewage being dumped into the ocean, it happens on a regular basis. The oceans are vast and can break down this vile liquid, but it still causes many adverse effects on marine life. Sewage or polluting substances flow through sewage, rivers, or drainages directly into the ocean. This is often how minerals and substances from mining camps find their way into the ocean. The release of other chemical nutrients into the ocean's ecosystem leads to reduction in oxygen levels, the decay of plant life, a severe decline in the quality of the sea water itself. As a result, all levels of oceanic life, plants and animals, are highly affected.

Domestic sewage and industrial effluents are released in the water courses in and around India in untreated or partially treated form. These, of course, add a mixture of pollutants which include, among others, certain toxic heavy metals and metalloids. The total volume of all discharges from the environs of Bombay is around 365 million tonnes (MT) per year. Similar discharges from the environs of Calcutta are around 350MT every year.

The birds hardly flock there and the fishes are dead, as no fauna can live in the toxic environment. The bay receives 64MT domestic sewage and 0. 9MT industrial effluents every year. These releases were initially untreated, but are now partially done by. The release of effluents of hydrocarbon origin has so heavily contaminated with the creek that it has become common practice to recover the oil by soaking Sorbents. Geochronology of sediments using 210Pb gave the maximum period since deterioration started at 54 years, which is roughly about the time when pumping of untreated sewage into the river started.

(3) Oil spills

The principal cause of marine pollution with oil is shipping. Traditionally shipping is considered to be "a polluting industry" . Ocean is polluted by oil on a daily basis from oil spills, routine shipping, run-offs and dumping. Oil spills make up about 12% of the oil that enters the ocean. The rest come from shipping travel, drains and dumping.

An oil spill from a tanker is a severe problem because there is such a huge quantity of oil being spilt into one place.

Oil spills causes a much localized problem but can be disastrous to local marine wildlife such as fish, birds and sea others.

Oil cannot dissolve in water and forms a thick sludge in the water. This suffocates fish, gets caught in the feathers of marine birds stopping them from flying and blocks light to photosynthetic aquatic plants.

Just an instance, On March 24, 1989, the Tanker Vessel Exxon Valdez ran aground 25 miles out of Valdez, Alaska. The impact tore open eight of the ship's eleven cargo tanks, spewing out 10. 8 million gallons of oil into Prince William Sound. Oil impacted hundreds of miles of pristine shoreline, inundating national forest and national park wilderness parcels. The spill wreaked havoc among sensitive coastal ecosystems, killing tens-of-thousands of waterfowl and other wildlife. The affected shoreline also contained significant archeological treasures.

The unprecedented attention of 1989 led to unprecedented statutory change. In 1990 a new oil pollution act was passed. The Oil Pollution Act of 1990 (OPA 90) was a landmark piece of legislation, going well beyond the scope of previous incremental provisions. The world's tanker fleet counts approximately 7000 vessels with cargo capacities between 76000 and 175000 tons. Usual shipping operations, especially transportation of oil by tankers and accidents, result in the dumping of around 600000 ~ 1750000 tons of oil into the ocean per year. Sea-bed activities on oil exploration and production constitute a relatively small part in the general amount of the pollution of marine environment with oil.

There are many harmful effects of oil spill on the environment. It kills animals and plants in the estuarine zone.

Oil kills the organisms that live on the beach if it settles there; it kills benthic organisms such as crabs if it settles on the ocean floor. Oil toxic algae, disturbs major food chains and results in the decrease of the production of edible oils. The oil that adheres to the body of the birds damages their flight and reduces the insulating property of their feathers, therefore causing the birds more vulnerable to cold.

(4) Non-point source

According to National Oceanic and Atmospheric Administration (NOAA), 80% of pollution to the marine environment comes from the land. One of the biggest sources is called non-point source pollution, which occurs as a result of runoff. Non-point source pollution includes many small sources, like septic tanks, cars, trucks, and boats, plus larger sources, such as farms, ranches, and forest areas. Millions of motor vehicle engines drop small amounts of oil each day onto roads and parking lots. Much of this, too, makes its way to the sea. Some water pollution actually starts as air pollution, which settles into waterways and oceans. Dirt can be a pollutant. Non-point source pollution can make river and ocean water unsafe for humans and wildlife. In some areas, this pollution is so bad that it causes beaches to be closed after rainstorms. Correcting the harmful effects of non- point source pollution is costly. Each year, millions of dollars are spent to restore and protect areas damaged or endangered by non-point source pollutants. NOAA works with several agencies to develop ways to control non- point source pollution. These agencies work together to monitor, assess, and limit non-point source pollution that may result naturally and by human actions.

NOAA's Coastal Zone Management Program is helping to create special non-point source pollution control plans for each coastal state participating in the program. When non-point source pollution does cause problems, NOAA scientists help track down the exact causes and find solutions.

选自：M. Vikas and G. S. Dwarakish/Aquatic Procedia 4 (2015) 381-388.

Vocabulary

detergents [dɪˈtədʒənt] *n.* 清洁剂

benthic organisms 水底生物

3.1.3　水净化

Water Purification

Waste water collected from municipalities and communities must ultimately be returned to receiving water or to the land or reused. The complex question facing the design engineer and public health officials is: What levels of treatment must be achieved in a given application–beyond those prescribed by discharge permits–to ensure protection of public health and the environment? The answer to this question requires detailed analyses or local conditions and needs application of scientific knowledge and engineering judgment based on past experience, and consideration of federal, state, and local regulations. In some cases, a detailed risk assessment may be required. An overview of wastewater treatment is provided in this section. The reuse and disposal of bios lids, vexing problems for some communities, are discussed in the following section.

Treatment methods

Methods of treatment in which the application of physical forces predominate are known as unit operations. At the present time, unit operations and processes are grouped together to provide various levels of treatment known as preliminary, primary, advanced primary, secondary (without or with nutrient removal) and advanced (or tertiary) treatment (shown in Table 3. 1). In preliminary treatment, gross solids such as large object, rags, and grit are removed that may damage equipment. In primary treatment, a physical operation, usually sedimentation, is used to remove the floating and settleable materials found in wastewater. For advanced primary treatment, chemicals are added to enhance the removal of suspended solids and to a lesser extent, dissolved solids. In secondary treatment, biological and chemical processes are used to remove most of the organic matter. In advanced treatment, additional combination of unit operations and processes are used to remove residual suspended solids and other constituents that are not reduced significantly by conventional secondary treatment. A listing of unit operations and processes used for the removal of major constituents found in wastewater and addressed in this text is presented in Table 3. 1.

Table 3. 1 The treatment level and description

Treatment level	Description
Prelimingary	Removal of wastewater constituents such as rogs, sticks, floatables, grit, and grease that may cause maintenance or operational problems with the treatment operations, processes, and ancillary systems
Primary	Removal of a portion of the suspended solids and organic matter from the wastewater
Advanced primary	Enhanced removal of suspended solids and organic matter from the wastewater
Secondary	Removal of biodegradable organic matter (in solution or suspension) and suspended solids
Secondary with nutrient removal	Removal of biodegradable organics, suspended solids, and nutrients (nitrogen, phosphorus, or both nitrogen and phosphorus)

Continued Table 3. 1

Treatment level	Description
Tertiary	Removal of residual suspended solids (after secondary treatment), usually by granular medium filtration or microscreens. Disinfection is also typically a part of tertiary treatment. Nutrient removal is often included in this definition
Advanced	Removal of dissolved and suspended materials remaining after normal biological treatment when required for various water reuse applications

About 20 years ago, biological nutrient removal (BNR)—for the removal of nitrogen and phosphorus—was viewed as an innovative process for advanced wastewater treatment. Because of the extensive research into the mechanisms of BNR, the advantages of its use, and the number of BNR systems that have been placed into operation, nutrient removal, for all practical purposes, has become a part of conventional waste water treatment. When compared to chemical treatment methods, BNR uses less chemical, reduces the production of waste solids, and has lower energy consumption.

Because of the importance of BNR in wastewater treatment, BNR is integrated into the discussion of theory, application, and design of biological treatment systems.

Land treatment processes, commonly termed "natural systems". combine physical, chemical, and biological treatment mechanisms and produce water with quality similar to or better than that from advanced wastewater treatment. Natural systems are not covered in this text as they are used mainly with small treatment systems; descriptions may be found in the predecessor edition of this text (Metcalf & Eddy, 1991) and in Crites and Tchobanoglous (1998) and Crites et al. (2000).

Current status

Up until the late 1980s, conventional secondary treatment was the most common method of treatment for the removal of BOD and TSS. In the United States, nutrient removal was used in special circumstances, such as in the Great Lakes area, Aorida, and the Chesapeake Bay, where sensitive nutrient-related water quality conditions were identified. Because of nutrient enrichment that has led to eutrophication and water quality degradation (due in part to point source discharges), nutrient removal processes have evolved and now are used extensively in other areas as well.

As a result of implementation of the Federal Water Pollution Control Act Amendments, significant data have been obtained on the numbers and types of wastewater facilities used and needed in accomplishing the goals of the program.

The municipal wastewater treatment enterprise is composed of over 16000 plants that are used to treat a total flow of about 1400 cubic meters persecond (m^3/s) [32,000 million gallons per day (M gal/d)]. Approximately 92 percent of the total existing flow is handled by plants having a capacity of 0.044m^3/s (1 million gallons per day and larger). Nearly one-half of the present design capacity is situated in plants providing greater than secondary treatment. Thus, the basic material presented in this text is directed toward the design of plants larger than 0.044m^3/s (1M gal/d) with the consideration that many new designs will provide treatment greater than secondary.

In the last 10 years, many plants have been designed using BNR effluent filtration has also been installed where the removal of residual suspended solids is required. Filtration is especially effective in improving the effectiveness of disinfection, especially for ultraviolet (UV) disinfection systems. Because (1) the removal of larger particles of suspended solids that harbor bacteria enhances the reduction in coliform bacteria and (2) the reduction of turbidity improves the transmittance of UV light effluent reuse systems, except for many that are used for agricultural irrigation, almost always employ filtration.

New directions and concerns

New directions and concerns in wastewater treatment are evident in various specific areas of wastewater treatment. The changing nature of the wastewater to be treated, emerging health and environmental concern, the problem of industrial wastes, and the impact of new regulations, all of which have been discussed previously, are among the most important. Further, other important concerns include:

(1) aging infrastructure,

(2) new methods of process analysis and control,

(3) treatment plant performance and reliability,

(4) wastewater disinfection,

(5) combined sewer overflows,

(6) impacts of storm water and sanitary overflows and nonpoint sources of pollution,

(7) separate treatment of return flows,

(8) odor control (shown in Fig. 3.2) and the control of VOC emission,

(9) retrofitting and upgrading wastewater treatment plants.

Fig. 3.2 Facilities used for chemical treatment of odors from treatment facilities

Aging infrastructure

Some of the problems that have to be addressed in the United States deal with renewal of the aging wastwater collection infrastructure and upgrading of treatment plants. Issues include repair and replacement of leaking and undersized sewers, control and treatment of overflows from sanitary and combined collection systems, control of nonpoint discharges, and upgrading treatment systems to achieve higher removal levels of specific constituents. Upgrading and retrofitting treatment plants is addressed later in this section.

Portions of the collection systems, particularly those in the older cities in the eastern and mid western United States, are older than the treatment plums. Sewers constructed of brick and vitrified clay with mortar joints, for example, are still used to carry sanitary wastewater and stormwater. Because of the age of the pipes and ancillary structure, the types of materials and methods of construction, and lack of repair. leakage is common. Leakage is in the form of both infiltration and inflow where water enters the collection system, and exfiltration where water leaves the pipe. In the former case, extraneous water has to be collected and treated, and oftentimes may overflow before treatment, especially during wet weather. In the latter case, exfiltration causes untreated waste water to enter the groundwater and/or migrate to nearby surface water bodies. It is interesting to note that while the standards for treatment have increased significantly, comparatively little or no attention has been focused on the discharge of untreated wastewater from sewers; through exfiltration. In the future, however, leaking sewer are expected to be come a major concern and will require correction.

摘自：Me tcalf & Eddy, In c. Wastewater Engineering Treatment and Reuse (Fourth Edition).

Vocabulary

sedimentation [ˌsedɪmenˈteɪʃn] *n.*　　沉淀

filtration [fɪlˈtreɪʃn] *n.*　　过滤

stormwater [stɔːmˈwɔːtə] *n.*　　雨水

retrofitting [ˈretrəfɪtɪŋ] *n.* 改装

treatment plants *n.* 水处理厂

disinfection [ˌdɪsɪnˈfekʃn] *n.* 消毒

3.2 大气污染控制技术

3.2.1 大气化学

Chemistry of the Atmosphere

The thin gaseous envelope that surrounds our planet is integral to the maintenance of life on earth. The composition of the atmosphere is predominately determined by biological processes acting in concert with physical and chemical change. Though the concentrations of the major atmospheric constituents oxygen and nitrogen remain the same, the concentration of trace species, which are key to many atmospheric processes are changing. It is becoming ap parent that man's activities are beginning to change the composition of the atmosphere over a range of scales, leading to, for example, increased acid deposition, local and regional ozone episodes, stratospheric ozone loss and potentially climate change. In this part, we will look at the fundamental chemistry of the atmosphere derived from observations and their rationalization.

In order to understand the chemistry of the atmosphere we need to be able to map the different regions of the atmosphere. The atmosphere can be conveniently classified into a number of different regions which are distinguished by different characteristics of the dynamical motions of the air the lowest region, from the earth's surface to the tropopause at a height of 10~15km, is termed the troposphere. The troposphere is the region of active weather systems which determine the climate at the surface of the earth. The part of the troposphere at the earth's surface, the planetary boundary layer, is that which is influenced on a daily basis by the underlying surface.

Above the troposphere lies the stratosphere, a quiescent region of the atmosphere where vertical transport of material is slow and radiative transfer of energy dominates. In this region lies the ozone layer which has an important property of absorbing ultraviolet (UV) radiation from the sun, which would otherwise be harmful to life on earth. The stratopause at approximately 50km altitude marks the boundary between the stratosphere and the mesosphere, which extends upwards to the mesopause at approximately 90km altitude. The mesosphere is region of large temperature extremes and strong turbulent motion in the atmosphere over large spatial scales.

Above the mesopause is a region characterized by a rapid rise in temperature, known as the thermosphere. In the thermosphere, the atmospheric gases, N_2 and O_2, are dissociated to a significant extent into atoms so the mean molecular mass of the atmospheric species falls. The pressure is low and thermal energies are significantly departed from the boltzmann equilibrium. Above 160km gravitational separation of the constituents becomes significant and atomic hydrogen atoms, the lightest natural species, moves to the top of the atmosphere. The other characteristic of the atmosphere from mesosphere upwards is that above 60km, ionisation is important. This region is called the ionosphere. It is subdivided into three regimes, the D, E and F region, characterized by the types of dominant photo ionisation.

With respect to atmospheric chemistry, though there is a great deal of interesting chemistry taking place higher up in the atmosphere, we shall focus in the main on the chemistry of the troposphere and stratosphere.

Sources of trace gases in the atmosphere

As previously described, the troposphere is the lowest region of the atmosphere extending from the earth's surface to the tropopause at 10 ~ 18km. About 90% of the total atmospheric mass resides in the troposphere and the greater part of the trace gas burden is found there. The troposphere is well mixed and its bulk composition is 78% N_2, 21% O_2, 1% Ar and 0.036% CO_2 with varying amounts of water vapor depending on temperature and altitude. The majority of the trace species found in the atmosphere is emitted into the troposphere from the surface and are subject to a complex series of chemical and physical transformations. Trace species emitted directly into the atmosphere are termed to have primary sources, e.g. trace gases such as SO_2, NO and CO. Those trace species formed as a product of chemical and/or physical transformation of primary pollutants in the atmosphere, e.g. ozone, are referred to as having secondary sources or being secondary species.

Emissions into the atmosphere are often broken down into broad categories of anthropogenic or "mass-made sources" and biogenic or natural sources with some gases also having geogenic sources. For the individual emission of a primary pollutant there are a number of factors that need to be taken into account in order to estimate the emission strength, these include the range and type of sources and the spatio- and temporal-distribution of the sources. Often these factors are compiled into the so-called emission inventories that combine the rate of emission of various sources with the number and type of each source and the time over which the emissions occur. It is clear from the data obtained, for example, SO_2 has strong sources from public power generation whereas ammonia has strong sources from agriculture. In essence, it has been apportioned spatially according to magnitude of each source category (e. g. road transport, combustion in energy production and transformation, solvent use). For example, the major road routes are clearly visible, showing NO_2 has a major automotive source. It is possible to scale the budgets of many trace gases to a global scale.

It is worth noting that there are a number of sources that do not occur within the bound ary layer (the decoupled lowest layer of the troposphere), such as lightning production of nitrogen oxides and a range of pollutants emitted from the combustion-taking place in aircraft engines 3. The non-surface sources often have a different chemical impact owing to their direct injection into the free troposphere (the part of the troposphere that overlays the boundary layer).

In summary, there are a range of trace species present in the atmosphere with a myriad of sources varying both spatially and temporally. It is the chemistry of the atmosphere that acts to transform the primary pollutants into simpler chemical species.

Initiation of photochemistry by light

Photodissociation of atmospheric molecules by solar radiation plays a fundamental role in the chemistry of the atmosphere. The photodissociation of trace species such as ozone and formal dehyde contributes to their removal from the atmosphere, but probably the most important role played by these photoprocesses is the generation of highly reactive atoms and radicals. Photo dissociation of trace species and the subsequent reaction of the photoproducts with other molecules is the prime initiator and driver for the bulk of atmospheric chemistry.

Vocabulary

tropopause [ˈtrɒpəʊpɔːz] *n.* 对流层顶
stratopause [ˈstretopɔz] *n.* 平流层顶
ionisation [ˌaɪənaɪˈzeɪʃən] *n.* 电离作用
radicals [ˈrædɪkl] *n.* 自由基
photodissociation [ˈfəʊtəʊdɪˌsəʊʃɪˈeɪʃən] *n.* 光离解

3.2.2 空气污染物及来源

Type and Sources of Air Pollutants

What Is Air Pollution? Air pollution is normally defined as air that contains one or more chemicals in high enough concentrations to harm humans, other animals, vegetation or materials. There are two major types of air pollutants. A primary air pollutant is a chemical added directly to the air that occurs in a harmful concentration. It can be a natural air component, such as carbon dioxide, that rises above its normal concentration, or something not usually found in the air, such as a lead compound emitted by cars burning leaded gasoline. A secondary air pollutant is a harmful chemical formed in the atmosphere through a chemical reaction among air components. Serious air pollution usually results over a city or other area that is emitting high levels of pollutants during a period of air stagnation. The geographic location of some heavily populated cities, such as Los Angeles and Mexico City, makes them particularly susceptible to frequent air stagnation and pollution buildup.

We must be careful about depending solely on concentration values in determining the severity air pollutants. By themselves, measured concentrations tell us nothing about the danger caused by pollutants, because threshold levels, synergy, and biological magnification are also determining factors. In addition, we run into the issue of conflicting views of what constitutes harm.

Major air pollutants following are the 11 major types of air pollutants:

(1) Carbon oxides, such as carbon monoxide (CO), and carbon

dioxide (CO_2).

(2) Sul fur oxides, such as sulfur dioxide (SO_2), and sulfur trioxide (SO_3).

(3) Nitrogen oxides, such as nitrous oxide (N_2O) nitric oxide (NO), and nitrogen dioxide (NO_2).

(4) Hydrocarbons organic compounds containing carbon and hydrogen, such as methane (CH_4), butane (C_4H_{10}), and benzene (C_6H_6).

(5) Photochemical oxidants ozone (O_3), PAN (a group of peroxyacylnitrates), and various aldehydes.

(6) Particulates (solid particles or liquid droplets suspended in air), such as smoke, dust soot, asbestos, metallic particles (such as lead, beryllium cadmium), oil, salt spray, and sulfate salts.

(7) Other inorganic compounds, such as asbestos, hydrogen fluoride (HF), hydrogen sulfide (H_2S), ammonia (NH_3), sulfur acid (H_2SO_4), and nitric acid (HNO_3).

(8) Other organic carbon-containing compounds, such as pesticides, herbicides, various alcohols, acids, and other chemicals.

(9) Radioactive substances tritium, radon, emissions from fossil fuel and nuclear power plants.

(10) Heat.

(11) Noise.

Table 3.2 summarizes the major sources of these pollutants.

Table 3.2　Major air pollutants and sources

Pollutants	Sources	Pollutants	Sources
Carbon oxides carbon monoxide (CO)	1. Forest fires and decaying organic matter, incomplete combustion of fossil fuels (about two-thirds of total emissions and other organic matter in cars and furnaces; 2. Cigarette smoke	Hydrocarbons	1. Incomplete combustion of fossil fuels in automobiles and furnaces; 2. Evaporation of industrial solvents and oil spills, tobacco smoke; 3. Forest fires; 4. Plant decay (about 85 percent of emissions)
Carbon dioxide (CO_2)	Natural aerobic respiration of living organisms burning of fossil fuel	Asbestos	1. Asbestos mining; 2. Spraying of fire proofing insulation in buildings; 3. Deterioration of brake linings
Sulfur oxides (SO_2 and SO_3)	1. Combustion of sulfur-containing coal and oil in homes, industries, and power plants; 2. Smelting of sulfur-containing ores; 3. Volcanic eruptions	Metals and metal compounds	1. Mining: industrial processes; 2. Coal burning; 3. Automobile exhaust
Particulates dust, soot and oil	1. Forest fires, wind erosion, and volcanic eruptions, coal burning: farming, mining construction, road building and other land-clearing activities; 2. Chemical reactions in the atmosphere; 3. Dust stirred up by automobiles; 4. Automobile exhaust; 5. Coal-burning electric power and industrial plants	Hydrogen sulfide (H_2S)	Chemical industry petroleum refining

Continued Table 3. 2

Pollutants	Sources	Pollutants	Sources
Pesticides and herbicides	1. Agriculture; 2. Forestry; 3. Mosquito control	Ammonia (NH_3)	Chemical industry petroleum refining
Nitrogen oxides (NO and NO_2)	1. High-temperature fuel combustion in motor vehicles and industrial and fossil fuel power plants; 2. Lightning	Sulfuric acid (H_2SO_4)	1. Reaction of sulfur trioxide and water vapor in atmosphere; 2. Chemical industry
Photochemical oxidants	Sunlight acting on hydrocarbons and nitrogen oxides	Nitric acid (HNO_3)	1. Reaction of sulfur trioxide and water vapor in atmosphere; 2. Chemical industry
Noise	Automobiles, airplanes, and trains industry construction	Ocher inorganic com pounds hydrogen fluoride (HF)	Petroleum refining glass etching aluminum and fertilizer production

Reading material

Type and Sources of Air Pollutants [II]

What are PCBs?

There are 209 possible chlorinated biphenyls, ranging in physical characteristics. The mono-and dichloro biphenyls (27323-18-8), (25512-42-9) are colorless crystalline compounds that when burned in air give rise to soot and hydrogen chloride. The most important products are trichlorobiphenyls, tetrachlorobiphenyl, pentachlorobiphenyl and/or hexachlorobiphenyls.

Chlorinated biphenyls are soluble in many organic solvents and in water only in the ppm range. Although chemically stable (including to oxygen of the air) they can be hydrolyzed to oxybiphenyls under extreme conditions forming toxic polychlorodibenzofurans. The PCB class of compounds received substantial attention and notoriety when in 1968, in Japan, accidental poisoning occurred by cooking rice in bran oil contaminated by PCBs over 1000 patients suffered from various morbid symptoms. A similar poisoning occurred in Taiwan in 1979. Causative agents were considered to be contaminated of PCBs such as poly chlorinated dibenzofurans that are secondarily formed during heating of PCBs congeners in commercial PCB mixtures and require a second look at PCB toxicity.

PCB regulation

Because of concerns regarding PCB's health effects and evidence of presence and persistence in the environment further manufacture of the chemical was banned under The 1976 Toxic Substance Control Act. PCB regulations provide deadlines for removal of most in-use capacitors and transformers containing PCBs and limit time for storage for disposal to one year.

EPA has allowed continued use of PCBs in electrical transformers and capacitors when the agency did not pose unreasonable risk. Capacitors, except those in isolated areas should have been removed by October 1988, and transformers of a certain size in or near commercial buildings should be removed by October 1990.

EPA regulations require that PCBs taken out of service be disposed of either by specially designed high temperature incinerators needed to break high concentrations of PCBs to harmless components or by alternate destruction methods approved by the agency. Oils contaminated with low concentrations (0.005%~0.05%) may be disposed by high efficiency boilers.

The limited number of incinerators approved for PCB incineration and the high cost of building additional incinerators have given incentive for alternate destruction methods. Alternate technologies must be capable of operating as effectively as EPA's incineration efficiency.

PCBs in electrical transformers

There were 304 million 1b of PCBs used as electrical fluid in approximately 150000 askarel (non-flammable electrical fluid) transformers in the United States. About 70000 PCB transformers are in or near commercial buildings that are open to the public. about 40000 of these transformers are owned by electrical utilities. Approximately 15000 of these transformers are used in the food and feed industry.

Utilities and other industries must maintain or dispose of approximately 150000 askarel-type transformers that may develop leaks. Each year, an estimated 317 askarel-type transformers can be expected to leak. Each will lose about 5.3 gal of PCBs.

Transformers classification

The EPA has three classification for transformers: PCB transformers contain 500 or more ppm PCBs, they must be inspected quarterly for leaks, and detailed records must be kept. No maintenance work involving removal of the core is allowed at the end of the transformers' useful life, it must be destroyed in an EPA-approved facility, or the transformer liquid must be incinerated and the carcass landfilled. The courts and the EPA have held the original transformer owner liable for leakage that may occur for as long as the carcass remains in the landfill.

PCB-contaminated transformers contain between 0.005% ~ 0.04% PCBs and require annual inspection. The rule concerning disposal, maintenance, and record keeping are less re strictive and less costly than those for PCB transformers Non-PCB transformers have less than 0.005% PCBs and are exempt from the burden some rules and requirements that apply to PCB and PCBs contaminated transformers. Non PCB transformers are granted favor under the Toxic Substances Control Act.

Analysis of PCBs

Analytical methods most frequently used for detecting chlorinated biphenyls are capillary column gas chromatography coupled with mass spectrometry in the MID (Multiple Ion Detection) mode and capillary column gas chromatography with ECD (Electron Capture Detector). HPLC (High Pressure Liquid Chromatography) and infrared spectroscopy are applicable to a limited extent. Summary PCB determinations are also possible, though not usual.

These require exhaustive chlorination and measurement of the decachlorbipheny content or dechlorination and subsequent biphenyl measurement.

Infrared tested on PCBs

Infrared thermal treatment technology was field tested for effectiveness on PCBs contaminated soil at two superfund sites during 1986. The transportable pilot system consisted of a primary furnace through which solid/semisolid wastes were conveyed on a wire mesh belt. The heat source was supplied by electric glow bars in lieu of gaseous fuels. Residence times and furnace temperatures could be controlled manually or automatically giving uniform ash. A secondary chamber, heated by electric or gaseous fuels, followed the primary furnace and provided temperatures in excess of 220F. The system was designed to comply with all RCRA and TSCA requirements.

3.2.3 大气污染防治新技术

New Technologies of Air Pollution Control

Biofiltration: an innovative air pollution control technology for VOC emissions

The concept of using microorganisms for the removal of environmentally undesirable compounds by biodegradation has been well established in the area of wastewater treatment for several decades. Not until recently, however, have biological technologies been seriously considered in the United States for the removal of pollutants from other environmental media. Moreover, while bioremediation techniques are now being applied successfully for the treatment of soil and groundwater contaminated by synthetic organics, at present there is very little practical experience with biological systems for the control of air contaminants among environmental professionals in the U. S. In fact, few environmental professionals in this country appear to be aware that "biofiltration", i. e, the biological removal of air contaminants from off-gas streams in a solid phase reactor, is now a well established air pollution control (APC) technology in several European countries, most notably The Netherlands and Germany.

In Europe, biofiltration has been used successfully to control odors, and both organic and inorganic air pollutants that are toxic to humans [air toxics, as well as volatile organic compounds (VOC) from a variety of industrial and public sector sources]. The development of biofiltration in West Germany, most of which took place in the late 1970s and the 1980s was brought about by a combination of increasingly stringent regulatory requirements and financial support from federal and state governments. The experiences in Europe have demonstrated that biofiltration has economic and other advantages over existing APC technologies particularly if applied to off-gas streams that contain only low concentrations (typically less than 0. 1% as methane) of air pollutants that are easily biodegraded.

The principal reasons why biofiltration is not presently well recognized in the U.S., and has been applied in only a few cases, appear to be a lack of regulatory programs, little governmental support for research and development, and lack of descriptions written in the English language. Specifically, regulatory programs in most U. S. states have not yet addressed, in a comprehensive manner, the control of air toxics, VOC and odors from smaller sources. Moreover, little financial support for investigating the applicability of biofiltration for these sources has been provided by government agencies. Finally, although several important papers on biofiltration have been published in English, most of the technical reports summarizing recent results, were published in Germany.

Despite these current obstacles, biofiltration is likely to find more widespread application in the U. S. in the near future. In addition to a few existing installations, several fullscale projects are currently in the planning stage or under construction. For example, the first large scale system for VOC control in California, a biofilter to treat ethanol emissions from an investment foundry in Los Angeles area is being planned with co-funding by the South Coast Air Quality Management District (SCAQMD). A detailed description of this system, and an analysis of its performance are provided elsewhere.

The major objective of the present paper is to provide a comprehensive review of important aspects of biofiltration in order to more widely disseminate about this innovative APC technology, and to encourage its implementation where appropriate in the U. S. Many of the more complex technical and engineering issues related to the development and use of bio filtration cannot be discussed in great depth here. However, we identify and summarize such issues, and refer to more detailed publication. We also note that, in addition to biofiltration other biological APC systems are now in use in Europe for the control of organic off-gas, including "bioscrubbers" and trickling filters. Various articles on these related technologies which are not discussed here, are available in other literature.

Suggestions to treat odorous off-gases by biological methods can be found in literature as early as 1923 when Bach discussed the basic concept of the control of H_2S emissions from sewage treatment plants. Reports on the application of this concept dating back to the 1950s were published in the U.S. and in West Germany. Pomeroy received U. S. patent No. 2, 793, 096 in 1957 for a soil bed concept and describes a successful soil bed installation in California. Around 1959 a soil bed was also installed at a municipal sewage treatment plant in Nuremberg, West Germany for the control of odors from an incoming sewer main.

In the U. S., the first systematic research on the biofiltration of H_2S was conducted by Carlson and Leiser in the early 1960s. Their work included the successful installation of several soil filters at a wastewater treatment plant near Seattle and demonstrated that biodegradation rather than sorption accounted for the odor removal.

During the following two decades, several researchers in the U. S. have further studied the soil bed concept and demonstrated its usefulness in several full scale applications. Much of the knowledge about the technology is owed to Hinrich Bohn who has investigated the theory and potential applications of soil beds for more than 15 years. Successful soil bed applications in the U. S. include the control of odors from rendering plants, and the destruction of propane and butane released from an aerosol can filling operation.

While soil beds have been shown to control certain types of odors and VOC efficiently and at fairly low capital and operating cost, their use in the U. S. has been limited by the low biodegradation capacity of soils and the correspondingly large space requirements for the beds. It is estimated that the total number of biofilter and soil bed installations in the U. S. and Canada is currently less than 50 and that they are predominantly used for odor control.

Reading material new technologies of air pollution control

Capital and operating costs associated with managing wastes produced by SO_2 emissions control equipment are important factors in evaluating and comparing alternative SO_2 control systems. Over the past eight years, EPRI has conducted a number of studies to provide utilities with cost information on waste management for conventional wet scrubbing. More recently a comprehensive investigation has been undertaken to assess waste management costs and issues for five alternate sulfur-reduction technologies: spray drying, atmospheric fluidized bed combustion, limestone furnace injection, dry-sodium injection, and advanced coal cleaning. For each of the five, studies have characterized waste products; developed engineering designs for effective waste handling, disposal, and/or utilization; and estimated waste management costs.

The first study, completed in late 1986, evaluated spray dryer wastes. Results showed that these wastes can be managed without excessive operating and economic problems for utilities or adverse environmental impacts. However, on a dollar-per-ton-disposed basis spray dryer waste management costs were found to be higher than those for either conventional fly ash or scrubber sludge alone. This finding indicates that cost estimates for new and retrofit spray dryer applications must be revised upward from those produced earlier by EPRI, under which waste management costs from all sulfur-reduction processes were assumed to be equal.

The process of a typical spray dryer waste management system involves five basic activities:

(1) Transfer of waste material from the spray dryer and particulate control device to a temporary storage facility.

(2) Storage.

(3) Conditioning to improve the material's handling characteristics (e.g., adding water to reduce fugitive dust emissions).

(4) Transportation to a disposal site or to a location where the material is utilized.

(5) For waste material not utilized, placement and containment at a disposal area.

Methodology and results

The spray dryer waste management study was conducted in four steps: characterizing spray dryer waste, surveying existing and planned spray dryer installations, developing conceptual designs and case studies for new and retrofit spray dryer installations, and evaluating the utilization potential of spray dryer waste.

Waste characterization

Waste material from seven utility spray dryer installations was analyzed to measure physical chemical, and leachate properties important in the design of a waste management system. Results show that while properties of spray dryer waste are generally similar to those of conventional fly ash, spray dryer waste is finer and more caustic, has a higher heat of hydration, and produces a more alkaline leachate. It can also become tacky when wet, and exhibits self-hardening properties similar to cement. Flowability test results indicate that normal relatively dry spray dryer waste is generally free to average flowing. With higher moisture contents, however, the material may set up and create serious storage problems. Flowability testing also indicates that spray dryer waste aerates easily and retains air once aerated.

Such characteristics are advantageous if controlled fluidized handling is used, but also indicate a potential for flooding, flushing, and flow rate limitation problems. (Flooding and flushing refer to conditions where an aerated bulk solid behaves like a fluid and flows uncontrollably through an outlet or feed mechanism.)

Survey of waste management systems

Data from 18 existing and planned spray dryer waste management systems were collected. Results indicate that waste management methods and equipment for these systems are similar to those used for conventional fly ash. Separate conveyors are typically used to transfer wastes from the spray dryer and particulate control device to a temporary storage silo. To transfer waste from the particulate collector, dilute-phase pressure pneumatic systems are most common, although pneumatic vacuum systems also are being used successfully. To transfer waste from the spray dryer, mechanical conveyers (primarily drag chain, bucket or screw-type) were initially installed in the existing facilities surveyed. However, all but five of these conveyors were converted to pressure or vacuum pneumatic systems due to operational problems caused by excessive wear and abrasion of parts. Most facilities practicing dry disposal methods condition the wastes with a small amount of water prior to transport by truck to an unlined landfill.

Conceptual designs and case studies

Based on results of the waste characterization and survey, and using the EPRI Technical Assessment Guide (TAG), spray dryer waste management conceptual designs were developed, and new and retrofit spray dryer waste management systems were designed for a hypothetical pulverized-coal-fired power plant burning low-ash western subbituminous coal.

For both the new and retrofit designs, collected waste was transferred to a common surge silo-spray dryer waste via a pneumatic system and from there to either a slurry preparation area for recycle or to a disposal silo via a dilute-phase pressure pneumatic conveyance system Waste withdrawn from the disposal silo is assumed to be conditioned with water and transported by truck to landfill For both the new and retrofit cases, the power plant was assumed to be a two-unit 1000-MW station. The new facility was assumed to be equipped with a fabric filter, whereas the existing station was assumed to be using an electrostatic precipitator at the time of the retrofit. Estimated total annual levelized costs for the waste management systems up to and including placement in an unlined landfill were $3029000 (0.53 mills/(kW · h) or $14.69/t) for the new plant, and $2760500(0.63 mills/(kW · h) or $17.40/t) for the retrofit installation.

Utilization

Based on spray dryer waste, physical properties, handling characteristics, chemical composition, environmental effects, and processing requirements, current fly ash utilization options were evaluated and ranked in terms of technical feasibility and marketability. In this analysis, the utilization potential for spray dryer wastes was found to be similar to that for C self-hardening ashes commonly found at plants in the midwestern Unite States of the options considered, the following seven applications were determined to be most attractive:

(1) Structural fill,

(2) Cement replacement,

(3) Stabilized road base,

(4) Synthetic aggregate,

(5) Lightweight aggregate,

(6) Mineral wool,

(7) Brick production.

3.3　环境影响

3.3.1　固体水的来源与类型

Sources and Types of Solid Waters

Since the beginning, humankind has been generating waste, be it the bones and other parts of animals they slaughter for their food or the wood they cut to make their carts. With the progress of civilization, the waste generated became of a more complex nature. At the end of the 19th century the industrial revolution saw the rise of the world of consumers. Not only did the air get more and more polluted but the earth itself became more polluted with the generation of nonbiodegradable solid waste. The increase in population and urbanization was also largely responsible for the increase in solid waste.

Each household generates garbage or waste day in and day out. Items that we no longer need or do not have any further use for fall in the category of waste, and we tend to throw them away. There are different types of solid waste depending on their source. Solid waste can be classified into three types depending on their source:

(1) Household waste is generally classified as municipal waste.

(2) Industrial waste as hazardous waste.

(3) Biomedical waste or hospital waste as infectious waste.

Municipal solid waste

Municipal solid waste consists of household waste, construction and demolition debris, sanitation residue, and waste from streets. This garbage is generated mainly from residential and commercial complexes. With rising urbanization and change in lifestyle and food habits, the amount of municipal solid waste has been increasing rapidly and its composition changing. In 1947 cities and towns in India generated an estimated 6 million tonnes of solid waste, in 1997 it was about 48 million tonnes. More than 25% of the municipal solid waste is not collected at all; 70% of the Indian cities lack adequate capacity to transport it and there are no sanitary landfills to dispose of the waste. The existing landfills are neither well equipped or well managed and are not lined properly to protect against contamination of soil and groundwater.

Over the last few years, the consumer market has grown rapidly leading to products being packed in cans, aluminum foils, plastics, and other such nonbiodegradable items that cause incalculable harm to the environment. In India, some municipal areas have banned the use of plastics and they seem to have achieved success. For example, today one will not see a single piece of plastic in the entire district of Ladakh where the local authorities imposed a ban on plastics in 1998. Other states should follow the example of this region and ban the use of items that cause harm to the environment. One positive note is that in many large cities, shops have begun packing items in reusable or biodegradable bags. Certain biodegradable items can also be composted and reused. In fact proper handling of the biodegradable waste will considerably lessen the burden of solid waste that each city has to tackle.

There are different categories of waste generated, each take their own time to degenerate (as illustrated in the Table 3.3).

Table 3.3　The type of litter and the approximate time of its degeneration

No.	Type of litter	Approximate time it takes to degenerate
1	Organic waste (vegetable, fruit peels, leftover foodstuff, ect.)	A week or two
2	Paper	10~30 days
3	Cotton cloth	2~5 months
4	Wood	10~15 years
5	Woolen items	1 years
6	Metals	100~500 years
7	Plastics	Undetermined
8	Glass	Undetermined

Hazardous waste

Industrial and hospital waste is considered hazardous as they may contain toxic substances. Certain types of household waste are also hazardous. Hazardous wastes could be highly toxic to humans, animals, and plants; are corrosive, highly inflammable, or explosive, and react when exposed to certain things e.g. gases. India generates around 7 million tonnes of hazardous wastes every year, most of which is concentrated in four states: Andhra Pradesh, Bihar, Uttar Pradesh, and Tamil Nadu Household waste that can be categorized as hazardous waste include old batteries, shoe polish paint tins, old medicines, and medicine bottles.

Hospital waste contaminated by chemicals used in hospitals is considered hazardous chemicals include formaldehyde and phenols, which are used as disinfectants, and mercury, which is in thermometers or equipment that measure blood pressure. Most hospitals in India do not have proper disposal facilities for these hazardous wastes.

In the industrial sector, the major generators of hazardous waste are the metal, chemical, paper, pesticide, dye, refining, and rubber goods industries. Direct exposure to chemicals in hazardous waste such as mercury and cyanide can be fatal.

Hospital waste

Hospital waste is generated during the diagnosis, treatment, or immunization of human beings or animals or in research activities in these fields or in the production or testing of biological. It may include wastes like sharps, soiled waste, disposables, anatomical waste, cultures, discarded medicines, chemical wastes, etc. These are in the form of disposable syringes, swabs, bandages, body fluids, human excreta, etc. This waste is highly infectious and can be a serious threat to human health if not managed in a scientific and discriminate manner. It has been roughly estimated that of the 4kg of waste generated in a hospital at least 1kg would be infected.

Surveys carried out by various agencies show that the health care establishments in India are not giving due attention to their waste management. After the notification of the Bio-medical Waste (Handling and Management) Rules, 1998, these establishments are slowly streamlining the process of waste segregation, collection, treatment, and disposal. Many of the larger hospitals have either installed the treatment facilities or are in the process of doing so.

Vocabulary

anatomical [ˌænəˈtɒmɪkl] *a.* 解剖的，解剖学的
cyanide [ˈsaɪənaɪd] *n.* [化学] 氰化物
demolition [ˌdeməˈlɪʃn] *n.* 破坏（pl.）废墟
lessen [ˈlesn] *vt.* 减轻，减少，轻视
vi. 变小，缩小，减少
immunization [ˌɪmjunaɪˈzeɪʃn] *n.* 免疫作用，有免疫力
landfill [ˈlændfɪl] *n.* 垃圾填埋场
formaldehyde [fɔːˈmældɪhaɪd] *n.* [化学] 甲醛
nonbiodegradable [ˌnaːnˌbaɪoudɪˈgreɪdəbl] *a.* 不可生物降解的
phenol [ˈfiːnɒl] *n.* [化学] 苯酚，石碳酸
sanitation [ˌsænɪˈteɪʃn] *n.* 卫生，卫生设施
slaughter [ˈslɔːtə(r)] *n.* 屠宰，（运动）大败
vt. 杀戮，屠宰，使（运动）

	大败
swab [swɒb] *n.*	[医] 药签，拖把
vt.	擦去，擦洗，拭抹，使用拖把
syringe [sɪ'rɪndʒ] *n.*	注射器，灌肠器，洒水器
vt.	注射，灌洗，冲洗

3.3.2 噪声控制

Noise Control

Industrial noise, which becomes increasingly evident with the increased number of industrial equipment, affects the health of the human hearing, digestive system, nervous system, endocrine system, etc. People have understood the harmful of noise pollution, and countries worldwide have formulated strict norms for industrial noise control. In these norms, the sound power and A-weighted noise levels are usually used to measure the noise, but they are not adequate to characterize the perception of a listener. The underlying concept of sound quality (SQ) is the accurate interference of human perception and was proposed by Blauert in 1994. The character of sound that relates to acceptance is called sound quality, which has played a large role in determining satisfaction. With the development of noise control technologies, sound quality research, which focuses on how people cognize, assess and improve noise, has gained attention, particularly in the fields of automobile, transportation and electric appliance industries worldwide.

(1) Automobile: Noise studies originated from the automobile industry in Europe and America in the mid-1980s. The main theoretical and experimental works on the human perception of sound quality were conducted by companies of AVL LIST, Honda, Delphi, Ford, GM, etc. Many automobile companies optimized the design of their products based on those research data.

(2) Transportation: Researchers also discussed the effects of the sound quality in aircraft, cabin, train and maglev trains.

(3) Electric appliances: The studies focused on air-conditioner, refrigerator, washing machine and mobile phone.

(4) Other SQ studies: Involving experiments and applications are introduced in Noise control can be classified into two types of methods: passive and active.

The passive noise control (PNC) method mainly reduces the noise by vibration absorption, sound absorption and sound insulation with damping materials by using the interaction between sound and materials, and the sound energy can be transformed into other forms of energy to reduce noise. The active noise control (ANC) method artificially adds a secondary source in the noise control process using Yaung's interference principle of sound wave to control the original noise. Compared with passive control, the active control methods have obvious benefits. First, the control system parameters can be targeted to design or change based on different characteristics of the noise. Second, the active control method has better control effect on low-frequency noise and effectively remedies the problem of low-frequency noise reduction effect. Finally, the active noise controller has the advantages of flexibility, low cost, and convenient installation; more importantly, it does not negatively affect the machine's structure and performance. The rapid development of large-scale integrated circuits and advancement of active control technologies have facilitated many successful implementations of ANC.

The active control method was proposed by Lueg in 1936 and applied for the process patent of acoustic-oscillation elimination in the United States; This patent is considered the starting point of the development of active noise control technology. In 1953, the first active noise control device, which was called "electronic sound absorber", was designed in the United States of America. This system consisted of a loudspeaker, an amplifier, and a microphone, and its target was to reduce the sound pressure level near the microphone. In the late 1950s, the acoustic field analysis technology was not mature, and the development of electronic technology was relatively slow. The active control technology was in a relatively quiet stage for a relatively long period of time until the 1980s. With the rapid development of digital signal processing and large-scale integrated circuit technology, the practical active noise control technology began to rapidly develop. Scientists in the United Kingdom first introduced the method of active noise control in automobiles and aircraft cabins. The least－mean－square (LMS) algorithm of channel filtering was used to study interior noise in Japan, and the active noise control model was established. In the United States of America, the detailed study and experiment of noise caused by engine vibration and road surface excitation were conducted by Jerome Couche, and the noise reduction of 6.5dB (A) was achieved in the range of 40～500Hz. Several prominent works on the development of ANC technology have been reported in the last three decades, such as the filtered-x least-mean-square (FxLMS) algorithm, genetic algorithm (GA), functional link artificial neural network (FLANN), simplified hyper－stable adaptive recursive filter (SHARF) algorithm and frequency selective least－mean－square (FSLMS) algorithm. In recent years, many research groups attempted to improve the noise sound quality using adaptive active noise control (AANC) methods. In Müller－BBM company, the experiment was performed on an AANC system, which was installed on a vehicle. The engineers found that the sound pressure level and loudness value (an objective parameter of SQ) of the interior noise significantly decreased. Spanish researchers conducted the engine noise active control in the lab, analyzed the psychoacoustic parameters, evaluated subjective evaluation results, and found that the reduction in sound pressure level did not necessarily reduce the annoyance of passengers to the engine noise, which was also related to the spectral characteristics of the noise.

3.3.3 环境影响评价摘要

Summary of Environmental Impact Assessment

An environmental impact assessment (EIA) is an assessment of the possible positive or negative impact that a proposed project may have on the environment, together consisting of the environmental, social and economic aspects.

The purpose of the assessment is to ensure that decision makers consider the ensuing environmental impacts when deciding whether to proceed with a project. The International Association for Impact Assessment (IAIA) defines an environmental impact assessment as "the process of identifying, predicting, evaluating and mitigating the biophysical, social, and other relevant effects of development proposals prior to major decisions being taken and commitments made". EIAs are unique in that they do not require adherence to a predetermined environmental outcome, but rather they require decision-makers to account for environmental values in their decisions and to justify those decisions in light of detailed environmental studies and public comments on the potential environmental impacts of the proposal.

EIAs began to be used in the 1960s as part of a rational decision-making process. It involved a technical evaluation that would lead to objective decision making. EIA was made legislation in the US in the National Environmental Policy Act (NEPA) of 1969. It has since evolved as it has been used increasingly in many countries around the world. As Jay et al. said, EIA is being used as a decision aiding tool rather than decision making tool There is growing dissent on the use of EIA as its influence on development decisions is limited and there is a view it is falling short of its full potential. There is a need for stronger foundation of EIA practice through training for practitioners, guidance on EIA practice and continuing research.

EIAs have often been criticized for having too narrow spatial and temporal scope. At present no procedure has been specified for determining a system boundary for the assessment. The system boundary refers to "the spatial and temporal boundary of the proposal's effects". This boundary is determined by the applicant and the lead assessor, but in practice, almost all EIAs address the direct, on-site effects alone.

However, as well as direct effects, developments cause a multitude of indirect effects through consumption of goods and services, production of building materials and machinery, additional land use for activities of various manufacturing and industrial services, mining of resources, etc. The indirect effects of developments are often an order of magnitude higher than the direct effects assessed by EIA. Large proposals such as airports or ship yards cause wide ranging national as well as international environmental effects, which should be taken into consideration during the decision-making process.

Broadening the scope of EIA can also benefit threatened species conservation. Instead of concentrating on the direct effects of a proposed project on its local environment, some EIAs used a landscape approach which focused on much broader relationships between the entire population of a species in question. As a result, an alternative that would cause least amount of negative effects to the population of that species as a whole, rather than the local subpopulation, can be identified and recommended by EIA.

There are various methods available to carry out EIAs, some are industry specific and some general methods.

Industrial products

Product environmental life cycle assessment (LCA) is used for identifying and measuring the impact on the environment of industrial products. These EIAs consider technological activities used for various stages of the product: extraction of raw material for the product and for ancillary materials and equipment, through the production and use of the product, right up to the disposal of the product, the ancillary equipment and material.

Genetically modified plants

There are specific methods available to perform EIAs of genetically modified plants. Some of the methods are risk assessment method for genetically modified plants (GMP-RAM, INOVA), eta.

Fuzzy arithmetic

EIA methods need specific parameters and variables to be measured to estimate values of impact indicators. However, many of the environment impact properties cannot be measured on a scale, e. g., landscape quality, lifestyle quality, social acceptance, etc. and moreover these indicators are very subjective. Thus to assess the impacts we may need to take the help of information from similar EIAs, expert criteria, sensitivity of affected population, etc. To treat this information, which is generally inaccurate, systematically, fuzzy arithmetic and approximate reasoning methods can be utilized. This is called as a fuzzy logic approach.

At the end of the project

An EIA should be followed by an audit. An EIA audit evaluates the performance of an EIA by comparing actual impacts to those that were predicted. The main objective of these audits is to make future EIAs more valid and effective. The two main considerations are: scientific–to check the accuracy of predictions and explain errors; management–to assess the success of mitigation in reducing impacts.

Some people believe that audits be performed as a rigorous scientific testing of the null hypotheses. While some believe in a simpler approach where you compare what actually occurred against the predictions in the EIA document. After an EIA, the precautionary and polluter pays principles may be applied to prevent, limit, or require strict liability or insurance coverage to a project, based on its likely harms. Environmental impact assessments are sometimes controversial.

Vocabulary

assessment [əˈsesmənt] *n.*	评估；（为征税对财产所作的）估价，被估定的金额
mitigate [ˈmɪtɪgeɪt] *vt. & vi.*	使镇静；使缓和，减轻（病情），平息（怒气）
prior to	在前，居先

参考文献

[1] Glasson J, Therivel R, Chadwick A J. Introduction to Environmental Impact Assessment: Principles and Procedures, Process, Practice, and Prospects [M]. London: Psychology Press, 1999.

[2] Jay S, Jones C, Slinn P, et al. Environmental impact assessment: Retrospect and prospect [J]. Environmental Impact Assessment Review, 2007, 27 (4): 287-300.

[3] Lenzen M, Murray S A, Korte B, et al. Environmental impact assessment including indirect effects: A case study using input-output analysis [J]. Environmental Impact Assessment Review, 2003, 23 (3): 263-282.

[4] Shepherd A, Ortolano L. Strategic environmental assessment for sustainable urban development [J]. Environmental Impact Assessment Review, 1996, 16 (4): 321-335.

[5] Jesus-Hitzschky K R E, Silveira J M F J. A proposed impact assessment method for genetically modified plants (AS-GMP Method) [J]. Environmental Impact Assessment Review, 2009, 29 (6): 348-358.